Globalization, Health, and the Environment

GLOBALIZATION AND THE ENVIRONMENT

This AltaMira series publishes new books about the global spread of environmental problems. Key themes addressed are the effects of cultural and economic globalization on the environment; the global institutions that regulate and change human relations with the environment; and the global nature of environmental governance, movements, and activism. The series will include detailed case studies, innovative multi-sited research, and theoretical questioning of the concepts of globalization and the environment. At the center of the series is an exploration of the multiple linkages that connect people, problems, and solutions at scales beyond the local and regional. The editors welcome works that cross boundaries of disciplines, methods, and locales, and which span scholarly and practical approaches.

SERIES EDITORS:

Richard Wilk, Department of Anthropology, 130 Student Building, Indiana University, Bloomington IN 47405 USA or wilkr@indiana.edu

Josiah Heyman, Department of Sociology & Anthropology, Old Main Building #109, University of Texas at El Paso, 500 West University Avenue, El Paso, TX 79968 USA or jmheyman@utep.edu

BOOKS IN THE SERIES

1. *Power of the Machine: Global Inequalities of Economy, Technology, and Environment,* by Alf Hornborg (2001)
2. *Confronting Environments: Local Environmental Understanding in a Globalizing World,* edited by James Carrier (2004)
3. *Communities and Conservation: Histories and Politics of Community-Based Natural Resource Management,* edited by J. Peter Brosius, Anna Lowenhaupt Tsing, and Charles Zerner (2005)
4. *Communities and Conservation: Histories and Politics of Community-Based Natural Resource Management,* edited by J. Peter Brosius, Anna Lowenhaupt Tsing, and Charles Zerner
5. *Globalization, Health, and the Environment: An Integrated Perspective,* edited by Greg Guest

Globalization, Health, and the Environment

An Integrated Perspective

EDITED BY
GREG GUEST

PRESS

A Division of
ROWMAN & LITTLEFIELD PUBLISHERS, INC.
Lanham • New York • Toronto • Oxford

AltaMira Press
A division of Rowman & Littlefield Publishers, Inc.
A wholly owned subsidiary of The Rowman & Littlefield Publishing Group, Inc.
4501 Forbes Boulevard, Suite 200
Lanham, MD 20706
www.altamirapress.com

PO Box 317, Oxford OX2 9RU, UK

British Library Cataloguing in Publication Information Available

Library of Congress Cataloging-in-Publication Data

Globalization, health, and the environment : an integrated perspective / [edited by] Greg
Guest.
    p. cm. — (Globalization and the environment)
    Includes bibliographical references and index.
    ISBN 0-7591-0580-4 (cloth : alk. paper) — ISBN 0-7591-0581-2 (pbk. : alk. paper)
    1. Environmental health. 2. World health. 3. Globalization—Health aspects. 4.
Globalization—Environmental aspects. I. Guest, Greg, 1963– II. Title. III. Series.
    RA565.G55 2005
    363.7—dc22
                                                                        2005007048

Printed in the United States of America

♾™ The paper used in this publication meets the minimum requirements of American
National Standard for Information Sciences—Permanence of Paper for Printed Library
Materials, ANSI/NISO Z39.48-1992.

Contents

Foreword

The dust raised by the lively debate about globalization and its consequences will take a while to settle. The process is complex and evolving. It has many facets, and the impacts on human societies are heterogeneous and often hard to attribute.

Globalization is essentially about human connection—via trade, investment, travel, the flow of ideas, shared cultural practices, political influences, and information technologies. There is, one senses, something inexorable about this broadly based connecting of peoples and cultures at a global scale. Indeed, viewed within a larger historical frame, there is a romantic dimension to it.

All members of the human species once cohabited within a relatively confined region of eastern Africa. Modern *Homo sapiens* originated around a quarter-million years ago, apparently in and around the Rift Valley—which runs across modern Ethiopia, Kenya, and Tanzania—by evolution from the upright-walking australopithecines and their early *Homo* species descendants. In those early millennia, that region of Africa would have had a carrying capacity that supported an estimated 50 thousand humans. Then, from around 100 thousand years ago, our ancestors began to disperse out of Africa, perhaps under stimulus of the oncoming glaciation and its impact on food species. Humans radiated into the eastern Mediterranean region, through southwest Asia, into East Asia and the Siberian region, across the Bering peninsula, into North America and, by 10 thousand to 15 thousand years ago, had extended southward into South America. Meanwhile, around 50 thousand years ago, some turned eastward from the Asian steppes and headed toward and into Europe, while others had pushed on around the southeast Asian shoreline and entered the north of Australia.

So, for most of the past one hundred millennia, we humans have been drifting apart, entering new environments, developing cultures attuned to local geography and ecology, forging distinctive languages, and, often, acquiring nuanced biological traits that optimized the fit between human organism and local environment. Complete isolation was not common, however. Neighboring tribes were in contact. Later, neighboring city-states warred with one another, and eventually great powers emerged in some parts of the world that were able to extend their influence over larger swaths of continents.

But the global reconnecting of humankind awaited the surge of European exploration, conquest, and imperialism, beginning around five hundred years ago. Here, then, were technologically advanced Europeans coming into unequal, fascinated, and sometimes violent contact with other long-settled peoples and cultures. European science, in those expansionist centuries, knew nothing of the grand story of human evolution and global dispersion. Rather, the world and its spectrum of oft-exotic peoples were seen as the set-piece product of the Divine Plan. Before the 19th century, there were no questions about "where from" and "when."

Despite these strong undercurrents over recent centuries, moving us increasingly toward a globalized world, the process has come upon us seemingly quite rapidly at the recent turn of the century. The accelerating pace of human technological change; the reach of modern capital investment, advertising, and trade; and the prominence of the Bretton Woods international instruments of economic development and financing in a new, postcolonial age have all converged in a process that we call "globalization." The freeing up of foreign exchange rates and the exponential increase in foreign direct investment facilitated by transnational agencies and instruments have given particular prominence to the "economic" dimension of globalization. This has been reinforced by the widely held, though simplistic, assumption that without conventional Western-style economic development, there can be no social advance, alleviation of poverty, improvement of health, and capacity to join the family of developed trading nations.

Over the past decade, however, there has been a spreading awareness across a widening constituency of interests that there is a substantial downside to this form of globalization. In particular, it has become obvious that the privileging of economic considerations, wherein "growth" is seen as an end in itself rather than as a means to realizing other human needs and values, has had many negative social and environmental consequences. For many, then, globalization has come to mean the seemingly unstoppable assembling of an economic juggernaut, particularly serving the interests of the already-rich nations. The World Trade Organization, spawned by the earlier General Agreement on Tariffs and Trade (GATT), has become a focal point of dissent. So long as world trade is driven by economic interests and profitability, without consideration of its impacts on the distribution of wealth, the conditions of production, the community's health, and the integrity of the natural environment (the true source of all new "wealth"), then this aspect of globalization is failing.

In this book, the particular focus is on how globalization impinges on the well-being and health of diverse human populations. Given the many dimensions of globalization, this is a useful framework within which to examine many health-related topics, including food supply and nutrition; patterns of infectious diseases; the rise of obesity in modern urban populations; the erosion of traditional medical practices and medicines; the intimate connections between land use and community morale and health; the mental and physical health consequences of enforced migration and population mobility; and the interplay between global, regional, and local human actions and well-being.

This wide-ranging volume also explores the relationship between the intensification of human economic and social activities, the rise of urbanized values and consumer behaviors, and the increasingly massive pressure that we humans are now collectively putting on the biosphere. Globalization per se is not necessarily *the* cause of global climate change, but many of the values and processes of economic and technological globalization, as currently practiced, have amplified the climate change process and other global environmental changes. As long as we have globalization processes driven primarily by markets, the profit motive, and national self-interest, we will continue to degrade our natural resource base.

The advent of the suite of global environmental changes that are now attracting our attention holds the potential for a final mighty irony for the human species. Having manifested a unique biological capacity to spread and occupy the whole world—and then, in recent centuries, to explore it and rediscover one another and consequently create international agencies of governance—we have now unwittingly embarked on an unsustainable pathway that is likely to seriously diminish the earth's life-supporting capacity over the next few decades. We have begun to notice. Can we now act?

There are, of course, solutions. We humans have an unusual capacity to foresee the future and estimate future risks. That said, however, we generally behave in ways that are true to the short-term imperatives of natural selection, of our inherited, evolved, human biology and psyche. After all, the struggle for survival throughout the natural world is, and always has been, necessarily conducted in the present tense. That is, actions are taken to survive today and tomorrow; the future will be dealt with similarly by our successful descendants, for whom it will be the "present."

So, we humans now face a supreme, unprecedented test of our capacity for enlightened collective action at a global level and in relation to human needs in the foreseeable future. The final chapter of this well-ordered, multiauthored volume offers us an agenda for achieving that rectification of globalization processes and ensuring that we move toward a sustainable path to the future. This is now an urgent task.

I have suggested in this foreword that there is something inexorable about the process of globalization. Although there are many aspects of the process that are detrimental to human interests, as well as many flavors that currently leave a nasty taste in the mouth, we must address the topic realistically. We have a widespread tendency to exalt the past, to mourn the loss of real or imagined simplicity and diversity. So we must ask ourselves, frankly, whether we really do wish to reattain the world of our grandparents or whether, instead, we should commit to working for a world for our grandchildren that is sustainable and supportive of happy and healthy human life—for all—and that preserves much of the cultural and environmental diversity of the current world.

We are currently at a sort of halfway house. We have moved from a world of somewhat isolated and independent tribes and populations to a world of generalized interdependence. The United Nations, the Olympic Games, the Internet, CNN television news, mobile phones, multilateral trading regimes, and so on—all are expressions of this ongoing connecting of human societies around the world. Some of this interdependence is positive; some is negative. The Palestinians and Israelis, for example, cannot avoid interacting with each other in today's world, given the reach of weapons, news media, and international politics. They are interdependent, albeit in a mostly negative way for the time being.

What is more generally missing, however, is the culminating step—the integration of human affairs and capacities. This is evident on many fronts. We have the means to eliminate the polio virus, but we lack the international stability and infrastructure to finish the job. How can we expect to deliver the antiretroviral therapy that will stop the HIV/AIDS virus in its tracks if we have the drugs and the best intentions but lack the professional capacity, political commitment, and infrastructure to distribute the drugs? Often the nation-state itself is an obstacle. Increasingly, in a globalized world we face complex, large-scale issues that cannot be addressed by a cacophony of self-interested nation-states. Consider the problems that we struggle to address in the realms of global climate change, the management of ocean fisheries, the global sales of cigarettes, the international narcotics trade, and so on.

Much of the thrust of this book is about the health consequences of globalization. In concluding, I note the obverse relationship, too. It is likely, for example, that infectious disease concerns will increasingly become a force for improved international integration. The experience with SARS (severe acute respiratory syndrome), in 2003, was that rapidity of international response—within a framework of agreed procedures and priorities and with increasing standardization and dissemination of data—facilitated the control of this potentially menacing viral disease.

We have come to better understand that in a world of intensified human economic activity, environmental encroachment, medical technology, and social contact networks, the opportunities for microbial adventurism—sometimes yielding new human infectious diseases—are much increased. If we are to live in such a world, globalized in multiple ways, then we must have the appropriately integrated structures of governance, international cooperation, and information sharing, leavened by mutual trust and respect for cultural diversity and enlightened by an understanding of how the human species must relate to the natural world.

Tony McMichael
National Centre for Epidemiology and Population Health
Australian National University

I

GLOBAL
TRANSFORMATIONS

1

Globalization, Health, and the Environment: An Introduction

Greg Guest and Eric C. Jones

While a German tourist savors a Coke and e-mails home from a mountain hut 16 thousand feet up Mount Everest, a hungry teenager consumes a Big Mac in Beijing, served with the standard McDonald's service and smile. Global Internet connections are growing rapidly and extend ever further into the remote reaches of the globe. Likewise, intercontinental air travel is accessible to millions of more tourists and businesspeople than ever before. Multinational corporations continue to grow and expand into the developing world, reaching ever further into once isolated markets.

These are all examples of globalization, a process that significantly affects the everyday lives of people around the world. This book is about the effect that globalization is having, and will likely continue to have, on our planet's ecology and on humans as a species. The chapters in this book demonstrate how globalization, health, and the environment are intricately interwoven, and while the authors may vary in their theoretical approaches, their case studies and perspectives form a collective engagement with the aspects of globalization that impinge on human health. In this introductory chapter, we present an overview of globalization, the global ecosystem, and world health to set the context for the chapters that follow.

GLOBALIZATION

Although a number of definitions of globalization exist, we begin with Anthony Giddens's broad view of globalization, which he defines as "the intensification

of world-wide social relations which link distant localities in such a way that local happenings are shaped by events occurring many miles away and vice versa"
(1990:64). The intensification to which Giddens refers is driven by a combination of interrelated factors—production, distribution, communication, finance,
and empire building. Under his broad definition, the beginnings of globalization can be documented as early as five thousand years ago (Gills and Gunder
Frank 1990) but can be seen to really take off in the 1500s, when the transport
of goods, people, and ideas between Europe, Africa, and the Americas became
an embedded part of European economies (Wallertsein 1974).

   With the burgeoning of capitalism and the subsequent onset of the Industrial
Revolution in the 18th century, technology flourished and human populations
grew, accelerating the rate of global intensification. In the last few decades, the
reach of globalization has become even broader and encroaches deeper into
local infrastructures, affecting the daily lives of people in previously isolated
communities around the world. While much has been written about the intensification of global interconnectedness, less has been written about the ways globalization affects health in an ecological context. To begin to formulate the relationships among globalization, health, and the environment, we first review
some potentially useful ways of thinking about this global interconnectedness,
including both the material aspects of life and the more psychological or symbolic aspects of life.

   Economic conditions are the aspect of globalization that has probably received the most attention. Here, globalization and its effects are seen as an economic reorganization of space and time. Michael Kearney (1995:549–551), for
example, contends that globalization is a change not just in the number and
location of connections but also in the amount of time and space in which
transactions occur. The world economy undergoes changes that require companies to achieve a greater number of sales of products in less time and greater
numbers of sales in smaller spaces (see Harvey 1989). Through competition
for investors' money and consumers' attention, companies are forced to seek
higher profits and lower costs. Making faster sales in smaller spaces is one way
to increase profits; employing cheap labor in developing countries is another.

   Opening up new markets is yet another way in which corporations expand
revenue streams. In many cases, this requires improved transportation facilities, thus enhancing the connectivity between local and global arenas. Such
connectivity typically brings an influx of people, money, and products into lo-

cal communities, which in turn affects identity formation (Kearney 1995; King 1991), local social structures (Schaeffer 2003), and economic conditions (Hutton and Giddens 2000). Greg Guest (2003), for example, shows how the construction of a simple road to an Ecuadorian fishing community paved the way for in-migration and adoption of new technology among subsistence fishers. New fishing gear and a connection to domestic and international demand for shrimp transformed the community significantly, affecting not only local fishing practices but also the local economic structure and the nature of local ecological knowledge (Guest 2002).

Globalization also affects how people think and, ultimately, how cultures evolve. Greater exchange of information, symbols, and ideas among once-distant cultures can foster changes in the ways that individuals view their world and in the ways that people relate to one another as individuals or groups (e.g., Appadurai 2001; Mudimbe-Boyi 2002; Tomlinson 1991).

The ways that individuals view their world and the ways that people relate to one another are in turn mediated by language, the embodiment of a culture; and the linguistic makeup of our planet has changed significantly in the past few centuries. At the turn of the 15th century, humans spoke an estimated 15 thousand languages. Over the last five hundred years, that number has dwindled to approximately 68 hundred, and of these, only a few are used in international communication. Some estimates project that the number of languages spoken in the world will be cut in half by the end of this century (Garry et al. 2001) and may be more threatened than species of birds or mammals (Sutherland 2003). Ironically, ethnic minorities frequently use globalized icons and symbols in new ways to express themselves and help promote the persistence of their culture (cf. Spicer 1984).

Another preeminent characteristic of culture in the modern period is nationalism and being a citizen in a nation-state. Citizenship normally has entailed the construal of rights to an individual by the government of the geographically bounded area (i.e., state) in which one lives, in addition to entailing the responsibilities demanded of the individual. But increased movement of people, information, capital, and goods has produced some new kinds of citizenship that may not be applicable to only one country or just to its borders (e.g., Portes 2000). New forms of citizenship may have more to do with practicing one's culture within a different society, immigrating as a minority, being an ecological citizen, consuming international goods and services, and relating to or visiting other societies or cultures around the globe (Urry 1999:315).

In summary, our everyday economics, ideas, and social interactions all are changed by, and yet also mediate in some way, the processes of globalization. We do not have the space in this introduction to discuss all of the intricacies of globalization, its history, or its underlying causes. These topics are covered more thoroughly in other works (Wallerstein 1974; Lechner and Boli 2000; Appadurai 2001; Lewellen 2002; Schaeffer 2003; Manning 1999; Kalb et al. 2000). Rather, we provide examples to show the breadth of impact that globalization is having on the human species and our planet and to emphasize the extent and rapidity with which globalization is occurring. We offer for contemplation a few figures that exemplify the rapid acceleration of global interconnectivity:

Mobility. Fifty million people were living outside their country of origin in 1989, but this number had grown to 125 million just a few years later (Kane 1995). International air passengers numbered 75 million in 1970 and 409 million in 1996 (Robbins 1999:219), a per capita increase of 255%. Total international tourist arrivals increased by over 50% between 1990 and 2000 (World Tourism Organization 2001), while the population increased by only 15%. By 2010, international tourism will involve at least one billion travelers each year, estimates the World Tourism Organization.

Trade. Since 1950 the global economy has grown more than 500% (United Nations Development Programme 1999), while the world's population has grown by only 132%. At the same time, international trade regulations have decreased: 35 countries were without exchange controls affecting imports in 1970, but this number had grown to 135 by 1997 (Wolf 2000:9). Between 1980 and 1998, annual global trade of cultural goods—such as literature, music, arts, cinema, photography, radio, and television—grew approximately 400% (United Nations Educational, Scientific, and Cultural Organization 2000), while the world's population grew by 33%.

Media. The number of television sets in the world increased from 192 million in 1965 to 1.157 billion in 1995 (United Nations Educational, Scientific, and Cultural Organization 1997), an overall increase of 502% worldwide and a per capita increase of 240%. While the reach of television and other news media is expanding, control over its ownership is becoming more concentrated. In 1983, about 50 media companies controlled more than half of all broadcast media, newspapers, magazines, video, radio, music, publishing, and film in the United States. But by 1993 this number had shrunk to 20. Today, fewer than 10 multinational media conglomerates control the majority of what Americans (and, to some extent, citizens of other countries) see and hear (Wellstone 2000).

Telephone and Internet. Telephone and cellular subscribers have increased by 184% from 1991 to 2001, according to estimates by the International Telecommunications Union, and transborder telephone calls increased from 3.2 billion in 1985 to 20.2 billion in 1996 (Robbins 1999:219), a per capita increase of 430%. Internet users are predicted to double between 2002 and 2005, to total nearly one billion (United Nations 2002), while the population is estimated to increase by only 3.5%.

Political economy. Governments may be losing influence over their economies to nongovernmental organizations and business corporations. The number of registered international nongovernmental organizations increased by 20% between 1990 and 2000 to 37 thousand (Anheier et al. 2001). Of the 100 largest economies in the world, 51 are corporations; only 49 are countries (Anderson and Cavanagh 2000). In addition, the ratio between the wealthiest and the poorest countries in terms of per capita income has grown from 11 to 1 in 1870, to 38 to 1 in 1960, to 52 to 1 in 1985 (Stephens et al. 2000). In 1988, the average income of the world's wealthiest 5% of people was 78 times that of the world's poorest 5%; five years later, this ratio had grown to 123 to 1 (World Bank 1999). The gap continues to widen.

Undoubtedly, we have seen increased intensity in antiglobalization movements in response to the expansion of global production, trade, and resource extraction. Counterglobalization efforts come in various forms and include such diverse movements as the World Social Forum, anti–World Trade Organization demonstrations, BUY LOCAL campaigns, indigenous battles against oil exploration or gold mining, and the antitechnology lifestyle of groups such as the Amish, to name a few (see McElroy, this volume). Globalization is thus of great interest to many people—be it positive, negative, or otherwise. But globalization is not just of concern to humans. It also weighs heavily on other species and their biophysical environments, a connection we discuss in the following section.

THE GLOBAL ECOSYSTEM

The environmental consequences of globalization have not escaped the notice of scholars, as evidenced by a recent emergence of edited volumes and monographs covering topics such as governance (Paehlke 2003), migration (Brah et al. 1999) and regional development (Rowntree et al. 1999). Other inquiries focus on the inequitable distribution of environmental and social damage imparted by global capitalism (Spaargaren et al. 2000; Paehlke 2003; Hornborg 2001; Kick and Jorgenson 2003). Regardless of which aspect of the problem one chooses to study, degradation of our biophysical environment owes to several common factors, including the extraction of resources, the transformation of the landscape through building, the manufacture of goods, and intensified transportation. Globalization entails the expansion of all of these processes.

To provide perspective, consider the life of an average T-shirt that one would purchase without thought in the United States. J. Ryan and A. Durning (1997) traced the process required to produce a T-shirt made from cotton and polyester and to bring it to market. Equipment, materials (direct and indirect), and financing and management associated with this product involved more than 25 countries. Processing the shirt itself occurred in seven different sites across four countries. Fabricating the polyester spewed out one-quarter of the fabric's weight in air pollution and 10 times its weight in carbon dioxide. Cotton production required a 14-square-foot patch of soil in Mississippi that was fumigated, treated with a soil sterilizer, and sprayed five times with pesticides. Cotton seeds were also dipped in fungicide, and a defoliant was applied to the crop just before harvesting. The raw cotton was processed and dyed in two textile mills, entailing multiple industrial chemicals and associated toxic waste. The fabric was subsequently shipped to Honduras, sewn into a T-shirt, and packaged in a polyethylene bag from Mexico with a rectangular cardboard sheet made from Georgia pine. The finished T-shirt was then shipped by freighter to Baltimore, by train to San Francisco, and finally by truck to Seattle, where it was purchased.

All of the activities in this process require energy and produce various types of pollution. Successful production and distribution of products such as the T-shirt in our example also rely on the existence of, and continued extension and intensification of, global infrastructure, transportation, and communications systems. The numbers presented in the previous section show fairly clearly that, indeed, such expansion is occurring and accelerating.

Keeping in mind the environmental impact that production of one simple T-shirt has, it is not difficult to comprehend the cumulative effect that global consumption is having on the earth's health. Current rates of industrial production require unsustainable exploitation of natural resources and produce significant amounts of pollution. Air pollution from automobile exhaust and industry, for example, chokes the world's cities and pours greenhouse gases into the atmosphere. Yearly carbon emissions increased 400% between 1900 and 2000 (McMichael and Paules 1999); the world has not seen such a rate of increase for 20 thousand years or more. Fossil-fuel consumption accounts for 70% to 90% of carbon dioxide emissions (Intergovernmental Panel on Climate Change 2001:7).

Land use change—particularly deforestation that supplies world markets with wood, paper, and agricultural products—accounts for the remaining carbon dioxide emissions. Overall, 60% of all temperate hardwood forests, 45% of tropical rain forests, 70% of tropical dry forests and 30% of confer forests have been sold for timber and paper (e.g., for textbooks!), converted to pasture or agriculture, or removed to make way for urban and commercial development (Noss and Peters 1995; Noss et al. 1995). Between 1980 and 1990 alone, the world's tropical forest cover decreased by 8% (Food and Agricultural Organization 1995). Compounding this predicament, global consumption of wood and paper products is projected to increase by 25% from 1996 to 2010 (Food and Agricultural Organization 1999:50).

Unsustainable exploit of land has caused 20% to 30% of the arable land to be "seriously damaged or lost" between 1950 and 2000 (McMichael 2001:314). Intensive production and urbanization have also led to soil degradation. Topsoil, for example, is eroding more rapidly than it forms on about one-third of the world's farmland, which means that each year the planet must feed 90 million more people with 26 billion fewer tons of topsoil (Miller 1994:33).

Co-occurring with these losses of arable land and forest habitat is the extinction of various plant and animal species. The earth is losing biodiversity at an estimated rate of 50 to 100 times the average that would be expected in the absence of human activity (Watson et al. 1998), and some researchers predict that as many as one-half of all animal and bird species could become extinct within two hundred to three hundred years (United Nations Environmental Programme 1995). The world's fisheries are also in a crisis, as marine and coastal ecosystems are increasingly stressed (McGoodwin 1990; United Nations Environmental Programme 1995). Humans rely on the resilience of all of these species and their biomes.

High levels of consumption also lead to unmanageable amounts of solid waste. In the United States alone, the world's largest producer of consumer waste, consumers throw away enough tires each year to encircle the earth three times, as well as about 2.5 million returnable bottles each *hour* (Miller 1994:15). A report from the United Nations Environment Programme poignantly outlines the relationship between consumption, global trade, and ecological instability:

> Continued poverty of the majority of the planet's inhabitants and the excessive consumption of the minority are the two major causes of environmental degradation. The present course is unsustainable and postponing action is no longer an option. . . . At a time when the developed world needs to cut back on per capita consumption, transnational businesses are engaged in efforts to create a giant global middle class. [Stephens et al. 2000:44–45]

WORLD HEALTH

Changes in ecosystems, like those described here, affect the future of human health, whether through disease, pollution, accident, war, or natural disaster. In fact, the World Health Organization (1997) estimates that poor environmental quality is directly responsible for about 25% of the world's burden of disease.

Globalization shapes how diseases are distributed, both socially and geographically (Farmer 1999; McMurray and Smith 2001; Armelagos and Harper, this volume). Many infectious diseases are on the rise worldwide and are related to globalization. Global travel, for example, played an important role in the early distribution of HIV. Although the virus has been documented in humans as early as the 1950s (Zhu et al. 1998), a Canadian flight attendant was identified as patient-zero for the ongoing HIV epidemic in the United States, which began in the early 1980s. Today, HIV infects approximately 16 thousand people a day worldwide and is now the world's fourth-largest cause of death (Stine 2002). SARS (severe acute respiratory syndrome) is yet another well-publicized example of the connection between global travel and infectious disease (see, e.g., Eyles and Consitt, this volume).

Malaria, yellow fever, Lyme disease, and a host of other vector-borne diseases are also reemerging around the world. Duane Gubler (1998) attributes this resurgence to a combination of factors. Urbanization, deforestation, changing agricultural practices (e.g., insecticide use and insects' resulting resistance), and an increase in the use of disposable consumer goods all provide more breeding areas for disease-carrying mosquitoes. In fact, a recent study in the Amazon found that every 1% increase in deforestation is accompanied by an 8% increase in the number of malaria-carrying mosquitoes (Pearson 2003). Air travel and changes in pubic health policy are other important variables in the epidemiology of these diseases (Gubler 1998). Anthropogenic climate change is yet another factor and is expected to further alter the distribution of vector-borne diseases, as well as reduce food supplies in the developing world (Lee et al. 2002; Patz et al. 1996).

Noncommunicable diseases such as cancer, diabetes, and cardiovascular diseases have also become a burgeoning global health problem and are linked to common risk factors—tobacco and alcohol use, unhealthy diet, physical inactivity, occupation hazards, and environmental carcinogens (Beaglehole and Yach 2003; Loewenson 2001). About 60% of the global burden of disease comprises noncommunicable diseases, and this proportion is expected to grow (World Health Organization 2002). As cultures in developing countries adopt the heavily promoted Western lifestyle, so too do they acquire the corresponding chronic diseases (see, e.g., McElroy, this volume). These cultures will not, however, have the same access to medical care and treatment as their counterparts in developed countries (e.g., Sitthi-amorn et al. 2001; Thankappan 2001). The chapters in this book by Ann McElroy and George Luber illustrate the relationship between culture (diet), health, and global processes.

Global processes also affect mental health. Finding work in a globalizing world increasingly demands a willingness to migrate. War and political turmoil and natural disasters associated with global warming also require moving. This mobility leads to splintered social networks and weaker support structures, both of which are related to depression (Palinkas et al. 1990; Lara et al. 1997). Chronic diseases, such as those discussed in the previous section, can also cause depression. It is therefore no surprise that depression is predicted to become the second-most burdensome disease category in the world by 2020 (Lopez 1997).

Not only are the incidence and distribution of health problems affected by globalization, but so is health policy (see, e.g., Lee et al. 2002). Facing stricter laws in many countries, tobacco companies are taking their products elsewhere and influencing policy through expensive public relations campaigns (Collin 2003). In another example, World Bank loan requirements forced Senegal to institute user fees for health care and reduce health care spending by 20% between 1981 and 1988 (Weissman 1990). As a result, the poor sought out traditional healers instead of seeking the services of public or private doctors or hospitals (Somerville 1997). In 10 of the 13 countries implementing World Bank structural adjustment programs, government spending on health, welfare, and education decreased between 1981 and 1990 (Stephens et al. 2000). The final two chapters in this book further discuss the interplay between global governance, public health, environmental degradation, and trade policy.

We do not wish to suggest that globalization has not benefited humankind's health. In many ways it has. Medical technology and improvements in health care have extended the length of the average human life (see, e.g., chapters by McElroy and Joseph, this volume): the average female in Japan today lives to an astounding 85 years of age (Oeppen and Vaupel 2002). Infant mortality is at its lowest point in human history (United Nations Children's Fund 1997); people are living longer nowadays than ever before (United Nations Population Division 1996); and the percentage of people over 65 will increase from 5% in 1950 to 15% in 2050 (United Nations Population Division 1998).

However, it may be that as a species we are *too* adaptive. Empirically, this is impossible to test, as the answer rests far in the future. But the point is that human demographic trends and health achievements associated with globalization—such as lower infant mortality and increasing population (see Joseph, this volume)—may bring increased risks for future fertility, such as pollution and constraints on food production and distribution. To be fair, the fertility rate has been decreasing in all regions of the world over the last 40 years. However, the sheer magnitude of men and women of childbearing age has caused the total number of births, and thus population, to increase. According to the UN's medium estimate, population growth resulting from this "population momentum" is expected to top out at around nine billion people in 2075 (United Nations Population Division 2003:7).

Even if this predicted demographic plateau occurs, sustaining a global population of nine billion will undoubtedly tax the global ecosystem and quality of life for many of the earth's inhabitants. Some scholars argue further that population growth offsets much of the progress made toward global health, such as access to clean water and sanitation (Haines and Kammen 2000). Currently, about one-third of the world's population live in countries whose water resources are stressed, and this number is expected to double by 2025 (United Nations 1997). Even seemingly positive developments are not without their costs. Longer life spans increase the prevalence of diseases associated with old age and increase the burden of caring for the elderly in a given society because of a lower ratio of working taxpayers to the elderly.

We must also consider that health care and many of the health benefits associated with technological development are not distributed equally between countries. Antiretroviral drugs, for example, that can extend the life of an HIV-infected person up to a decade or more are often not available, or affordable, in many of the countries hardest hit by AIDS. In developing countries, the aspiring middle class is now experiencing obesity and its corresponding health problems (McLellan 2002), and the poor in these countries suffer from malnutrition (McMichael 2001:236).[1]

Whether globalization has had an overall positive or negative effect on human health is debatable (see, e.g., Cornia 2001; Dollar 2001; Feachem 2001; Lee et al. 2002). While the details of this complex discussion are well beyond the scope of this chapter, the situation is perhaps best characterized by McMichael and Beaglehole, in an article that appeared in the *Lancet*:

Economic globalization—entailing deregulated trade and investment—is a mixed blessing for health. Economic growth and the dissemination of technologies have widely enhanced life expectancy. However, aspects of globalisation are jeopardizing health by eroding social and environmental conditions, exacerbating the rich-poor gap, and disseminating consumerism. [2000:495]

We would like to stress at this point that the relationship between globalization and health is not exclusively a one-way process in which the former shapes the latter. The reverse can also be true. Health and disease can, and often do, facilitate global connectivity. They often serve as focal points for mobilizing international efforts, such as scientific inquiry, policy (see, e.g., Eyles and Consitt, this volume), and political activism (see, e.g., Epstein and Guest, this volume). Such mobilization and sharing of ideas between cultures and across physical space can in turn affect future health outcomes.

Regardless of how globalization and health are viewed, two things are certain: first, globalization is destabilizing in terms of its impacts on local environmental conditions, diet, health, and demography; and, second, global processes exert tremendous influence over the distribution of disease and the patterns in human health, hitting hardest those populations most marginalized and with the fewest resources. The chapters of this book examine these generalizations by using holistic frameworks to make connections between health, globalization, and the environment. They are designed to illustrate the intricacies of the interaction between these three processes.

Part I examines technological change over long periods and its relationship to global political–economic processes and health outcomes. The two general processes of concern are the production and trade of industrial goods and the human demographic patterns closely tied to agriculture and food production. Socioeconomic class plays a large part in mediating the incidence of the various health problems brought about by these two distinct yet interrelated processes.

Looking at human disease within a macrohistorical lens, George Armelagos and Kristin Harper, in the next chapter, describe the three major "epidemiological transitions" in human history. They illustrate how the shift from foraging to agriculture about ten thousand years ago marked the first transition, characterized by the emergence of infectious disease, nutritional deficiency, and social inequalities. The second epidemiological transition— associated with public health measures, medicine, and improved nutrition during the last century—has resulted in a decline in infectious disease and a rise in chronic and degenerative disease. Armelagos and Harper suggest that the human species is on the eve of a third epidemiological transition, in which new infectious diseases and a reemergence of known, but now resistant, infectious diseases are becoming increasingly problematic.

In chapter 3, Thomas Leatherman shows how worldwide patterns of health are intricately linked to changing biophysical and social environments. Leatherman discusses environmental health problems in the Third World, such as armed conflicts and violence; population dispersals and the creation of large numbers of internal and external refugee groups; environmental degradation and pollution; and the capitalist transformations of agrarian peoples. These problems result in poor nutrition and lowered immunities, plus exposure to disease pathogens and toxins. The chapter concludes with notes about the sorts of problems that might dominate research on global processes of environment and health in the future.

The chapters forming part II are case studies and deal with cultural adaptation by indigenous populations whose health-related behaviors and beliefs have been altered by global forces. David Casagrande, for example, documents in chapter 4 how global processes have influenced migration patterns of Tzeltal Mayans and the way in which some migrants have redefined their relations with a critical environmental health resource—medicinal plants. Specifically, and in contrast to conventional academic wisdom, Casagrande argues that knowledge about medicinal plants in the migrant communities has not necessarily been eroded but commodified. With integration into the global economy, what used to be widely shared "traditional" knowledge is now becoming the province of a few privileged Tzeltal.

In chapter 5, McElroy documents the effects of globalization on Inuit communities in the Canadian Arctic. She provides an ethnohistory of the Canadian Inuit and documents how national and international policies have disrupted the health ecology of residents—especially the elderly—of the newly formed territory of Nunavut. Traditional subsistence practices and nutritional health have all been altered. McElroy's research also illustrates the intricate relationship between subsistence patterns, diet, and Inuit identity.

In chapter 6, Luber reports a link between a Mesoamerican folk illness called "second hair" illness and changes in land use among Mexico's Mixe and Tzeltal peoples. Luber suggests that the shift from a subsistence-based economy to a monetized economy driven by global economic forces has altered dietary patterns to those favoring high-status, low-quality foods over the low-status, high-quality foods typical of the traditional subsistence diet.

Part III, "Population Dynamics," deals more explicitly with the association between globalization and demography. John Eyles and Nicole Consitt discuss in chapter 7 the local, national, and international responses to global forces that act as environmental "exposures" in Canada. They emphasize infectious disease (mad cow disease, tuberculosis, SARS) and the political responses to such threats. Eyles and Consitt point out the rapidity with which these diseases can spread in a world so well interconnected, and they discuss the role that the media and other information sources have in the control of such diseases.

Urbanization is a distinct phenomenon of globalization and is occurring throughout every country of the world. Mary Anne Alabanza Akers and Timothy Akers discuss this trend in chapter 8 and, employing a case study from the Philippines, illustrate how urban migration is paralleled by the rise of an informal economic sector found in the hidden landscapes of urban streets, corners, sidewalks, and alleyways. Their case study poignantly describes the health hazards associated with working in the shadows of an urban environment, such as road hazards, environmental contaminants, and exposure to violence.

Suzanne Joseph, in chapter 9, introduces us to the reproductive world of the Bedouins of the Bekaa Valley, Lebanon, and the impacts of transnational and local political–ecological forces on Bedouin microdemographic behavior. In particular, she shows how the decline of pastoralism as a means of livelihood and the increasing reliance on agriculture have led to a historical rise in fertility among the Bedouins (although their fertility has since dropped). She further documents how Bedouins' access to natal and health care facilities in Lebanon, adequate public sanitation, and the availability of high-quality weaning foods in the form of sheep and goat milk have combined to produce low infant–child mortality, as well as adequate nutrition. It is difficult to predict the future trajectory of Bedouin health; however, as Bedouins reduce their reliance on pastoralism and become more dependent on sharecropping and wage labor, they risk increasing impoverization and all of its associated negative health implications.

Finally, part IV shifts the frame of analysis to the policy arena. In chapter 10, Linda Whiteford and Beverly Hill's account of dengue fever in Cuba and the Dominican Republic provides insight into the reasons why public health policies might succeed or fail. Contrasting dengue-control programs in these two Caribbean countries, they highlight key aspects of effective control programs and corresponding health outcomes. Interestingly, Cuba, long isolated from the global political–economic community, has proven to implement environmental modification and subsequent control of the mosquito-transmitted dengue more effectively than its Latin American counterpart has.

Discussing the status of human and environmental health can cast a "doom and gloom" view of our humanity's future on this planet. However, it behooves us to move beyond simple descriptions of dismal scenarios or projections and to strive for a better understanding of the underlying political–economic, sociocultural, and historical forces that are most responsible for the human health and environmental problems we now face. Only in this way can we offer effective policy interventions or formulate models for human social and political action.

Paul Epstein and Greg Guest end the book with a forward-looking chapter, one that offers possible solutions and, ideally, stimulates similar attempts among the book's readers. After examining global economic policy, the environmental degradation of our planet, and the inequitable distribution of disease burden among the world's population, they conclude that equity and environmental preservation must be the guiding principles of any future policy (see Brecher et al. 2000 for some concrete suggestions on how to make this happen). They further advocate that this "new economic order" be based on a more inclusive mandate, one that includes not only politicians but also corporate leaders (particularly from the financial sector), labor unions/organizations, nongovernmental organizations, UN agencies, and representatives from groups that are most vulnerable to the adverse effects of unmanaged economic growth and environmental destruction.

As you, the reader, engage and contemplate the contents of this book, we ask that you, too, maintain a solution-oriented ember in the back of your mind. It is all too easy to either eschew responsibility for human health and environmental issues or fall back on a negative fatalism about the human condition. Perhaps what is most needed is an integrated perspective, alongside creative thinking, to guide globalization toward alternative futures.

NOTE

1. We should point out that health care is not necessarily distributed equitably within countries. Poverty and social marginalization among certain groups in the United States, for instance, impede adequate access to health care. Lee Warner and colleagues (2001), for example, documented 157 cases of congenital syphilis—an easily preventable disease—over a three-year period among children of low-income women in an urban southeastern hospital, illustrating how easy it is to fall through the cracks of medical care in one of the world's wealthiest nations.

REFERENCES

Anderson, Sarah, and John Cavanagh
2000   Top 200: The Rise of Corporate Global Power. Washington, DC: Institute for Policy Studies.

Anheier, Helmut, Marlies Glasius, and Mary Kaldor, eds.
2001   Global Civil Society 2001. New York: Oxford University Press.

Appadurai, Arjun, ed.
2001   Globalization. Durham, NC: Duke University Press.

Beaglehole, R. and D. Yach
2003   Globalisation and the Prevention and Control of Non-Communicable
        Disease: The Neglected Chronic Diseases of Adults. Lancet 362:903–908.

Brah A., M. Hickman, and M. Mac an Ghaille, eds.
1999   Global Futures: Migration, Environment, and Globalization. New York:
        St. Martin's Press.

Brecher, Jeremy, Tim Costello, and Brendan Smith
2000   Globalization from Below: The Power of Solidarity. Cambridge, MA:
        South End Press.

Collin, J.
2003   Think Global, Smoke Local: Transnational Tobacco Companies and
        Cognitive Globalization. In Health Impacts of Globalization: Towards
        Global Governance. Kelley Lee, ed. Pp. 61–82. Houndsmills, Hampshire,
        UK: Palgrave.

Cornia, Giovanni A.
2001   Globalization and Health: Results and Options. Bulletin of the World
        Health Organization 79:834–841.

Dollar, David
2001   Is Globalization Good for Your Health? Bulletin of the World Health
        Organization 79:827–833.

Farmer, P.
1999   Infections and Inequalities: The Modern Plagues. Berkeley: University of
        California Press.

Food and Agricultural Organization
1995   Forest Resources Assessment 1990—Global Synthesis. Forestry Paper, 124.
        Rome: Food and Agricultural Organization.
1999   State of the World's Forests. Rome: Food and Agricultural Organization.

Feachem, Richard G.
2001   Globalisation Is Good for Your Health, Mostly. British Medical Journal
        323:504–506.

Garry, Jane, Carl Rubino, Alice Faber, and Robert French, eds.
2001   Facts about the World's Languages: An Encyclopedia of the World's Major
        Languages, Past and Present. Bronx, NY: H. W. Wilson.

Giddens, Anthony
1990   The Consequences of Modernity. Stanford, CA: Stanford University.

Gills, Barry K., and Andre Gunder Frank
1990   The Culmination of Accumulation: Theses and Research Agenda for 5,000
        Years of World System History. Dialectical Anthropology 15:19–42.

Gubler, Duane
1998   Resurgent Vector-Borne Diseases as a Global Health Problem. Emerging
        Infectious Diseases 4:442–450.

Guest, Greg
2002   Market Integration and the Distribution of Ecological Knowledge within
        an Ecuadorian Fishing Community. Journal of Ecological Anthropology
        6:38–49.
2003   Fishing Behavior and Decision-Making in an Ecuadorian Community:
        A Scaled Approach. Human Ecology 31:611–644.

Haines, Andrew, and Daniel Kammen
2000   Sustainable Energy and Health. Global Change and Human Health 1:78–87.

Harvey, D.
1989   The Condition of Postmodernity: An Inquiry into the Origins of Culture
        Change. Cambridge: Blackwell.

Hornborg, A.
2001   The Power of the Machine: Global Inequalities of Economy, Technology,
        and Environment. Walnut Creek, CA: AltaMira.

Hutton, W., and A. Giddens, eds.
2000   Global Capitalism. New York: New Press
        Intergovernmental Panel on Climate Change
2001   Climate Change 2001: The Scientific Basis. Cambridge: Cambridge
        University Press.

Kalb, D., M. Van der Land, R. Staring, B. van Steenbergen, and N. Wilterdink, eds.
2000   The Ends of Globalization: Bringing Society Back. Boulder, CO: Rowman
        & Littlefield.

Kane, Hal
1995   The Hour of Departure: Forces That Create Refugees and Migrants.
        Worldwatch Paper, 125. Washington, DC: Worldwatch Institute.

Kearney, Michael
1995   The Local and the Global: The Anthropology of Globalization and
        Transnationalism. Annual Review of Anthropology 24:547–565.

Kick, E., and A. Jorgenson, eds.
2003   Globalization and the Environment. Theme issue, Journal of
        World-Systems Research 9(2).

King, A. D., ed.
1991   Culture, Globalization and the World System: Contemporary Conditions
        for the Representation of Identity. Binghamton: State University of New
        York Press.

Lara, M. E., J. Leader, and D. N. Klein
1997   The Association between Social Support and Course of Depression: Is It
        Confounded with Personality? Journal of Abnormal Psychology 106:478–482.

Lechner, Frank J., and John Boli
2000   The Globalization Reader. Malden, MA: Blackwell.

Lee, Kelley, Tony McMichael, Colin Butler, Mike Ahern, and David Bradley
2002   Global Change and Health—the Good, the Bad and the Evidence. Global
        Change and Human Health 3:16–19.

Lewellen, T.
2002   The Anthropology of Globalization: Cultural Anthropology Enters the 21st
        Century. Westport, CT: Begin and Garvey.

Loewenson, Rene
2001   Globalization and Occupational Health: A Perspective from Southern
        Africa. Bulletin of the World Health Organization 79:863–868.

Lopez, Murray A.
1997   Alternative Projections of Mortality and Disability by Cause 1990–2020:
        Global Burden of Disease Study. Lancet 349:1498–1504.

Manning, Susan, ed.
1999   Special Issue on Globalization. Theme issue, Journal of World-Systems
        Research 5(2).

McGoodwin, J. R.
1990   Crisis in the World's Fisheries: People, Problems, and Policies. Stanford,
        CA: Stanford University Press.

McLellan, Faith
2002   Obesity Rising to Alarming Levels around the World. Lancet 359:1412.

McMichael, A. J.
2001   Human Frontiers, Environments and Disease: Past Pasterns, Uncertain
        Futures. Cambridge: Cambridge University Press

McMichael, A. J., and R. Beaglehole
2000   The Changing Global Context of Public Health. Lancet 356:495–499.

McMichael, A. J., and J. W. Paules
1999   Human Numbers, Environment, Sustainability and Health. British Medical Journal 319:977–980.

McMurray, C., and R. Smith, eds.
2001   Diseases of Globalization: Socioeconomic Transitions and Health. Sterling, VA: Earthscan Publications.

Miller, G. Tyler, Jr.
1994   Environment: Problems and Solutions. Belmont, CA: Wadsworth.

Mudimbe-Boyi, M., ed.
2002   Beyond Dichotomies: Histories, Identities, Cultures, and the Challenge of Globalization. Albany: State University of New York Press.

Noss, R. F., E. T. LaRoe, and J. M. Scott
1995   Endangered Ecosystems of the United States: A Preliminary Assessment of Loss and Degradation. Biological Report, 28. Washington, DC: National Biological Service.

Noss, R. F., and R. L. Peters
1995   Endangered Ecosystems: A Status Report on America's Vanishing Habitat and Wildlife. Washington, DC: Defenders of Wildlife.

Oeppen, Jim, and James A. Vaupel
2002   Broken Limits to Life Expectancy. Science 296 (May 10):1030–1031.

Paehlke, Robert
2003   Democracy's Dilemma: Environment, Social Equity, and the Global Economy. Cambridge, MA: MIT Press.

Palinkas, L. A., D. L. Wingard, and E. Barrett-Connor
1990   The Biocultural Context of Social Networks and Depression among the Elderly. Social Science and Medicine 30:441–447.

Patz, Jonathon, Paul Epstein, Thomas Burke, and John Balbus
1996   Global Climate Change and Emerging Infectious Diseases. Journal of the American Medical Association 275:217–223.

Pearson, Helen
2003   Lost Forest Fuels Malaria: Conservation and Medicine Collide in the Jungle. Nature Science Update. Electronic document, www.nature.com/nsu/031124/031124-12.html, accessed November 28, 2003.

Portes, Alejandro
2000   Globalization from Below: The Rise of Transnational Communities. *In* The
        Ends of Globalization: Bringing Society Back In. D. Kalb, M. Van der Land,
        R. Staring, B. van Steenbergen, and N Wilterdink, eds. Pp. 253–270.
        Boulder, CO: Rowman & Littlefield.

Robbins, Richard H.
1999   Global Problems and the Culture of Capitalism. Boston: Allyn and Bacon.

Rowntree, L., M. Lewis, M. Price, and W. Wyckoff
1999   Diversity amid Globalization: World Regions, Environment, and
        Development. New York: Prentice Hall College Division.

Ryan, J., and A. Durning
1997   Stuff: The Secret Lives of Everyday Things. Seattle: Northwest Environment
        Watch.

Schaeffer, R.
2003   Understanding Globalization: The Social Consequences of Political,
        Economic, and Environmental Change. Lanham, MD: Rowman &
        Littlefield.

Sitthi-amorn, Chitr, Ratana Somrongthong, and Watana Janjaroen
2001   Some Health Implications of Globalization in Thailand. Bulletin of the
        World Health Organization 79:889–890.

Somerville, C.
1997   Reaction and Resistance: Confronting Economic Crisis, Structural
        Adjustment, and Devaluation in Dakar, Senegal. *In* Global Survival in the
        Black Diaspora: The New Urban Challenge. C. Green, ed. Pp. 15–41.
        Albany: State University of New York Press.

Spaargaren, G., A. Mol, and F. Buttel, eds.
2000   Environment and Global Modernity. Thousand Oaks, CA: Sage.

Spicer, Edward Holland
1984   Pascua, a Yaqui Village in Arizona. Tucson: University of Arizona Press.
[1940]

Stephens, Carolyn, Simon Lewin, Giovanni Leonardi, Miguel San Sebastian Chasco,
    and Richard Shaw
2000   Health, Sustainability and Equity. Global Change and Human Health
        1(1):44–58.

Stine, Gerald
2002   AIDS Update 2002. Upper Saddle River, NJ: Prentice Hall.

Sutherland, William
2003   Parallel Extinction Risk and Global Distribution of Languages and Species.
         Nature 423:276–279.

Thankappan, K. R.
2001   Some Health Implications of Globalization in Kerala, India. Bulletin of the
         World Health Organization 79:892–893.

Tomlinson, John
1991   Cultural Imperialism: A Critical Introduction. London: Pinter.

United Nations
1997   Comprehensive Assessment of the Freshwater Resources of the
         World. Commission on Sustainable Development. New York:
         United Nations / World Meteorological Organization / Stockholm
         Environment Institute
2002   E-Commerce and Development Report, 2002. New York: United Nations.

United Nations Children's Fund
1997   State of the World's Children. New York: United Nations Children's Fund.

United Nations Development Programme
1999   Human Development Report 1999. New York: Oxford University Press.

United Nations Educational, Scientific, and Cultural Organization
1997   Statistical Yearbook 1997. Paris: United Nations Educational, Scientific, and
         Cultural Organization.
2000   Study on International Flows of Cultural Goods, 1980–1998. Paris:
         United Nations Educational, Scientific, and Cultural Organization.

United Nations Environment Programme
1995   Global Diversity Assessment. Cambridge: Cambridge University Press.
1999   Global Environment Outlook. Nairobi, Kenya: Earthscan.

United Nations Population Division
1996   Demographic Indicators, 1950–2050: The 1996 Revision. New York:
         United Nations.
1998   World Population Prospects: The 1998 Revision. New York:
         United Nations.

2003   World Population in 2300. Proceedings of the United Nations Expert
          Meeting on World Population in 2300. New York: United Nations
          Population Division.

Urry, J.
1999   Globalisation and Citizenship. Journal of World-Systems Research
          5(2):311–324.

Wallerstein, Immanuel
1974   The Modern World System. New York: Academic Press.

Warner, Lee, Roger Rochat, Ronald Fichtner, Barbara Stoll, Lawrence Nathan, and
     Kathleen Toomey
2001   Missed Opportunities for Congenital Syphilis Prevention in an Urban
          Southeastern Hospital. Sexually Transmitted Diseases 28:92–98.

Watson, R. T., J. A. Dixon, S. P. Hamburg, A. C. Janetos, and R. H. Moss
1998   Protecting Our Planet, Securing Our Future: Linkages among Global
          Environmental Issues and Human Needs. Washington, DC: United Nations
          Environment Program / National Aeronautics and Space Administration /
          World Bank.

Weissman, S. R.
1990   Structural Adjustment in Africa: Insights from the Experiences of Ghana
          and Senegal. World Development 18(12):1621–1635.

Wellstone, Paul
2000   Introduction. In It's the Media, Stupid. Robert McChesney and John
          Nichols, eds. Pp. 15–20. New York: Seven Stories Press.

Wolf, Martin
2000   Why This Hatred of the Market. In The Globalization Reader.
          Frank J. Lechner and John Boli, eds. Pp. 9–12. Malden, MA: Blackwell.

World Bank
1999   Inequality: Trends and Prospects. Washington, DC: World Bank.

World Health Organization
1997   Health and Environment in Sustainable Development. Geneva: World
          Health Organization.
2002   The World Health Report, 2002. Geneva: World Health Organization.

World Tourism Organization, ed.
2001   Tourism 2020 Vision: Global Forecast and Profiles of Market Segments.
         Lanham, MD: Bernan.

Zhu, Tuofu, Bette Korber, Andre Nahmias, Edward Hooper, Paul Sharp and David Ho.
1998   An African HIV-1 Sequence from 1959 and Implications for the Origin of
         the Epidemic. Nature 391(6667):594.

2

Disease Globalization in the Third Epidemiological Transition

George J. Armelagos and Kristin N. Harper

The global nature of emerging disease has been described by the possibility of an Ethiopian sneezing and a Californian catching the cold. While this example may seem to exaggerate the rapidity of disease transmission in a shrinking world, one aspect of the remark is understated. A person who contracts a communicable disease in Addis Abba, Ethiopia, can be in Atlanta, Georgia, in less than a day, able to transmit a pathogen more deadly than the rhinovirus, which causes the common cold. Global disease ecology has experienced, and continues to experience, profound and rapid change.

One of the key objectives of this chapter is to demonstrate the role that emerging disease played in the adaptation and evolution of past human populations. There is a misplaced tendency to assume that emerging diseases are a recent phenomenon. Even the Institute of Medicine's definition of emerging diseases focuses on the "new, reemerging, or drug-resistant infections whose incidence in humans has increased within the past decades or whose incidence threatens to increase in the near future" (Hughes 2001). The emphasis on the immediacy of emerging diseases belies the fact that they have evolutionary depth. We recast emerging disease as an ongoing evolutionary process that extends back to the Neolithic period, an era characterized by the advent of agriculture and domestication of plants and animals.

The second concept that plays a key role in our analysis is the issue of globalization. In the popular literature, globalization has become an oft-used term for the "inexorable integration of markets, nation-states and technologies to a degree never witnessed before—in a way that is enabling individuals, corporations and nation-states to reach round the world farther, faster, deeper and cheaper than ever before" (Friedman 1999:7). The origins of globalization date as far back as five thousand years ago, but the process intensified in 16th-century Europe, as the transport of goods with Africa and the Americas increased and became a fundamental part of the European economy. In *Plagues and People*, W. H. McNeill (1976) describes a process in which civilizations "digest" the populations they encounter, as disease vectors clear the paths in advance, allowing for easy access and uncontested expansion. He describes the confluence of the civilized disease pools of Eurasia that formed from 500 B.C.E. until C.E. 1200. This was followed by the development of the Mongol Empire from C.E. 1200 to 1500, which resulted in the global spread of the plague bacillus. The technological developments from C.E. 1500 to 1700 led to the transoceanic exchange that culminated in a link between Old World diseases (i.e., from Europe, Asia, and Africa) and those from the New World (the Americas). Globalization, then, has been a process at work throughout history.

We argue that both disease emergence and globalization have considerable historic depth and are inextricably related. Using the concept of epidemiological transition, we attempt to meet three goals. First, we contextualize the concept of emerging disease as a feature of the last ten thousand years of our evolutionary history. Second, we place the globalization process within a historical context that extends over a thousand years. Third, we examine how globalization has shaped the pattern of emerging disease and how emerging diseases in turn affect globalization.

EPIDEMIOLOGICAL TRANSITION

Abdel Omran (1971) models a pattern of human history that moved through three disease stages. The earliest humans lived in an "age of pestilence and famine," which gave way to an "age of receding pandemics," culminating in "the age of degenerative and man-made diseases." Omran (1971, 1977, 1983) believes that the process of disease evolution will progress to a state in which infectious diseases are eradicated. With the elimination of early mortality due to infectious disease, the average age of individuals in the population will increase, and a rise in degenerative, chronic, "man-made" diseases will follow.

While Omran (1975, 1977, 1982, 1983) presents his model of epidemiological transition as a comprehensive analysis of human history, abundant bioarcheological evidence suggests that human populations underwent an earlier disease transition, one that he does not describe. Omran's scenario reflects an earlier conception of the Paleolithic as a disease-ridden era (or the part of the Stone Age extending from two million years before the end of the last ice age, roughly 8500 B.C.E., in which humans lived as hunter–gatherers); and he implies that infectious disease was a major burden experienced by all populations predating the modern era. The conventional wisdom of the time accepted the traditional Hobbesian view that Paleolithic hunter–gatherers lived a life "solitary, poor, nasty, brutish, and short" (*Leviathan*, i.xiii.9). It is now believed, however, that the shift from foraging (hunting–gathering) to primary food production (agriculture) resulted in an increase in infectious disease and nutritional deficiency. In this chapter, we broaden Omran's version of epidemiological transition to include this alternative view of the Paleolithic as a relatively disease-free era.

Such an expanded epidemiological model (Armelagos and Barnes 1999; Armelagos 1997; Armelagos et al. 1996; Barnes et al. 1999; Barrett et al. 1998) encompasses three transitions. While a distinct pattern of disease emerged among our Paleolithic ancestors (Desowitz 1980; Lambrecht 1967, 1980, 1985), their mobility, small population size, and low density precluded infectious disease from asserting a major evolutionary impact (Cockburn 1967a, 1971; Polgar 1964). The first epidemiological transition occurred ten thousand years ago with the subsistence shift from foraging to agriculture. A dramatic increase in infectious diseases resulted from the subsequent explosion in population size and density, the domestication of animals that served as sources of disease vectors, and the advent of sedentary populations (Armelagos 1990, 1991). This dramatic transition to primary food production has been dubbed the "Neolithic revolution" and is marked by the appearance of social stratification, which exaggerated differential risk to disease within a population (Armelagos and Brown 2002; Steckels and Rose 2002). The emergence of disease that marks the first epidemiological transition accelerates as agricultural land use intensifies and spreads to new regions.

The second epidemiological transition represents Omran's original disease transition. It is a period in which the development of medical practices, improved nutrition, and public health measures resulted in a decline in infectious disease (McKeown 1976). As average age of the population increased with the reduced threat of infectious disease, the population began to experience a rise in age-related chronic disease (Johansson 1991, 1992).

Enter the third epidemiological transition, characterized by the reemergence of infectious diseases previously thought to be under control (Kombe and Darrow 2001; McMichael 2000), many of which are antibiotic resistant (Georgopapadakou 2002; Mah and Memish 2000), in addition to the rapid emergence of a number of "new" diseases. The existence of pathogens resistant to multiple antibiotics portends the possibility that we are living in the eve of the antibiotic era. During our lifetime, it is possible that many pathogens resistant to all antibiotics will emerge. Furthermore, the third epidemiological transition is characterized by a transportation system that globalizes the infection process (Waters 2001). There are no points on the habitable earth that are more than 30 hours apart by air transportation. The transmission of pathogens is becoming so rapid and so extensive that we are now said to be connected by a virtual "viral superhighway" (Armelagos 1998).

PALEOLITHIC BASELINE

The disease ecology of a globalized world in which infectious diseases may be quickly spread between continents contrasts with the Paleolithic populations, in which infectious diseases had little impact on relatively isolated populations. This is not meant to imply that the Paleolithic was a disease-free Eden. An array of parasites would have affected our earliest ancestors (Cockburn 1967a, 1967b; Hasegawa 1999; Stevens et al. 1998). Early hominids were exposed to parasites deemed "heirloom species" (Sprent 1962, 1969a, 1969b) because they infested our anthropoid ancestors, coevolving with them as hominids emerged. Lice, *Pediculus humanus,* is one example of an heirloom species. Lice have existed as ectoparasites since the beginning of the Oligocene, about 33.7 million years ago, long before man walked the earth (Laird 1989), and head lice (*Pediculus humanus capitis*) were undoubtedly part of the early hominid's pests. Recent genomic analysis of lice (Kittler et al. 2003) suggests that the body lice (*Pediculus humanus humanus*) that now live in woven fabric differentiated from head lice about 70 thousand years ago, when clothing presumably became a part of human wardrobe. Other heirloom species include pinworms, *Enterobius vermicularis,* and other internal parasites that commonly infect modern humans, such as salmonella and staphylococci (Cockburn 1967a, 1967b).

In contrast to heirloom species, parasites whose relationships with humans are long-standing, "souvenir" species are occasionally picked up along the course of a hominid's daily activity. Such diseases are zoonoses, whose primary hosts are nonhuman animals and which only incidentally infect humans. Human zoonoses may be passed through animal bites (Weber and Rutala 1999), insect bites, and preparing and consuming contaminated flesh (Slifko et al. 2000). Sleeping sickness, tetanus, scrub typhus, relapsing fever, trichinosis, tularemia, avian or ichthyic tuberculosis, leptospirosis, and schistosomiasis are among the zoonotic diseases that likely afflicted earlier hunter–gatherers (Cockburn 1971).

Speculation on other zoonotic diseases in the Paleolithic proves more difficult. For example, the *Anopheles* mosquito, the vector for malaria, had adapted to the canopy environment of forests by the Miocene era (a period about 25 million years ago characterized by the emergence of grazing animals) and therefore would have been present in the Paleolithic, millions of years later. Even so, the threat malaria posed to early hominids would have been minor, since hominid populations were sparse. Because the major blood source for the *Anopheles* was nonhuman primates, the mosquito may have adapted to grassy woodlands but still not have been a threat to early hominids. It was only with the advent of slash-and-burn (swidden) agriculture, which creates a breeding ground for the mosquito near human settlements, that malaria became a constant threat. The increase in population size and density as well as the presence of a sedentary and dependable pool of human hosts would have allowed the mosquitoes to shift from nonhuman primates to hominids as a food source. Analysis of the frequency of the sickle-cell gene suggests that the trait is a relatively recent response to malaria (Livingstone 1958, 1971 1983), suggesting that if malaria was contracted by early hominids, it would have been in isolated incidents.

Additional recent genetic analyses support the contention that malaria had little impact on Paleolithic foragers and arose only after the development of swidden agriculture. The independent "A" and "Med" mutations in glucose-6-phosphate dehydrogenase, human genetic variations that provide an individual with some protection against malaria, arose no longer than ten thousand years ago (Tishkoff et al. 2001). Analysis of the *Plasmodium falciparum* genome supports this data, indicating a great expansion of the pathogen less than ten thousand years ago (Joy et al. 2003).

The range of the earliest hominids, who walked the earth more than four million years ago, was probably restricted to a tropical grassy woodland environment, limiting the type of pathogens that could be potential disease agents. Our human ancestors would have found the extensive areas of African savannah south of the Serengeti plains uninhabitable because of biting tsetse flies bearing *Trypanosoma brucei*, the causative agent of sleeping sickness (Dicke 1932; Lambrecht 1964, 1967, 1980, 1985). Although humans carry an antitrypanosome factor in their blood that prevents infection by many types of trypanosome (Zimmer 2001), we remain susceptible to infection by *T. brucei*. As hominids moved into new ecological niches, the pattern of trypanosome infection would have changed, intensifying enough to become a serious health threat only when humans began to move out of the tropical woodlands and colonize the African plains, where tsetse flies make their home.

Recent research has provided evidence for a zoonotic disease not previously thought to have posed a threat to early hominids. The three modern taeniid tapeworms that now parasitize humans were thought to have originated in the Neolithic with the domestication of animals. Genomic and biogeographic evidence now suggests that the tapeworms originated as human parasites much earlier, in the Paleolithic (Hoberg et al. 2001; Hoberg et al. 2000). E. P. Hoberg and colleagues claim that the three species of the host-specific taeniid tapeworms, *Taenia saginata*, *T. asiatica*, and *T. solium*, show extensive genetic diversity, which indicates that a species separation occurred between 780,000 and 1.71 million years ago, long before animal domestication. Hoberg and colleagues claim that this is evidence that a regular pattern of scavenging and hunting (Shipman 2002) resulted in the emergence of a human-specific tapeworm on three separate occasions, humans only later infecting their domestic animals with these tapeworms (Hoberg et al. 2001).

Small population size would have precluded many of the bacterial and viral infections that we are familiar with today. The diseases missing from the Paleolithic include the contagious diseases that require a high population density to be maintained in a population, such as influenza, measles, mumps, and smallpox. F. M. Burnet (1962) argues that there would have been few viruses infecting early hominids. T. A. Cockburn (1967b), however, claims that nonhuman primates would have been a source of viral infections in early hominids.

THE FIRST EPIDEMIOLOGICAL TRANSITION: AGRICULTURAL SOCIETIES

The earliest evidence of primary food production is found in the Old World. Five centers arose independently about ten thousand years ago, each domesticating different cultigens (Figure 2.1). These five centers include Mesopotamia, where barley and wheat were domesticated; sub-Saharan Africa, a center of millet and plantain production; Southeast Asia, relying on rice; a millet-based center in northern China; and southern China, for rice agriculture. The Tigris–Euphrates was the site of significant settlements by seven thousand years ago. A thousand years later, centralized governments utilizing extensive irrigation systems emerged. Centers in the New World originated later and were based on the domestication of maize in Mesoamerica and potatoes in South America. In the Valley of Mexico, there were well-established settlements by 3500 B.P., and by C.E. 1 these settlements showed extensive complexity. It is important to note that centralized polities are inherently hierarchical, with classes having differential access to resources, an important factor governing disease susceptibility.

The emergence of disease during the Neolithic was influenced by marked ecological changes following the shift to primary food production. In sedentary habitats, the accumulation of human waste represents an ideal source for disease transmission. Unlike hunting–gathering groups, which move frequently and make frequent forays away from the base camp, agriculturalists are permanently tied to the land. Often, human waste deposits are situated near the source of potable water, becoming an even greater source of infection.

The life of an individual inhabiting an early agricultural civilization was rife with disease threats. The domestication of animals fostered continuous contact between humans and disease vectors. Goats, sheep, cattle, pigs, fowl, and unwanted domestic animals such as rodents and sparrows developed permanent habitats in and around human dwellings and became sources of zoonotic diseases. Animal products such as milk, hair, and skin, as well as the dust raised by the animals, could transmit anthrax, Q fever, brucellosis, and tuberculosis. In addition, breaking the sod during cultivation exposes agriculturists to insect bites and diseases such as scrub typhus (Audy 1961). As discussed, F. B. Livingstone (1958) shows that slash-and-burn agriculture in West Africa exposed populations to *Anopheles gambiae*, the mosquito vector for *Plasmodium falciparum*, which causes malaria. The combination of disruptive environmental farming practices and the presence of domestic animals also

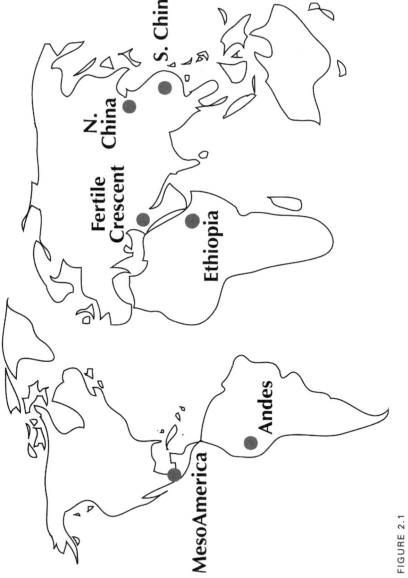

FIGURE 2.1
Independent origins of agriculture in the ancient world.

increased human contact with arthropod vectors carrying yellow fever, try-panosomiasis, and filariasis. Some vectors went so far as to develop dependent relationships on human habitats, the best example of which is *Aedes aegypti* (the vector for yellow fever and dengue), which breeds in artificial containers. Various agricultural practices, such as irrigation and the use of feces as fertil-izer, increased contact with nonvector parasites such as *Schistosomal cercariae* and intestinal flukes (Cockburn 1971).

Finally, food type and storage must be considered as a factor in the chang-ing disease ecology of the first epidemiological transition. The shift from a varied, well-balanced diet to one that contained fewer types of food some-times resulted in dietary deficiencies. Food was stored in large quantities and widely distributed, probably resulting in outbreaks of food poisoning from pathogens and toxins on a scale not previously encountered (Audy 1961). M. N. Cohen and G. J. Armelagos (1984) edited a volume containing a num-ber of studies that show a decline in health following the Neolithic. The com-bination of a complex society, increasing divisions of class, epidemic disease, and dietary insufficiencies no doubt added mental stress to the list of illnesses.

First Epidemiological Transition: Urban Development and Disease

Sedentary life and population growth are historically associated with agri-cultural development. The growth of urban centers and populations large enough to maintain many diseases in endemic form is a recent development in human history. In the Near East, large cities were established 67 hundred years ago. Urban centers at Memphis (Egypt) reached 30 thousand souls by 3100 B.C.E.; Ur (Babylonia) reached 65 thousand inhabitants by 2030 B.C.E.; and Babylon had a population of 200 thousand by 612 B.C.E. (Chandler 1987). Set-tlements of this size increased the already difficult problem of removing hu-man wastes and delivering uncontaminated water to inhabitants. Cholera, which is transmitted by contaminated water, was a potential problem. Diseases such as typhus (carried by lice) and the plague bacillus (transmitted by fleas or through the respiratory tract) could be spread from person to person. Viral dis-eases such as measles, mumps, chicken pox, and smallpox could be spread like-wise. Cockburn (1967a) estimates that populations of one million would be necessary to maintain measles as an endemic disease, while others (Black et al. 1974) suggest that a population of only two hundred thousand would do. Francis L. Black and colleagues (1974) argue that a population size of only one thousand people is needed to sustain chicken pox as an endemic disease. Because an endemic disease in one population could be the source of a serious epidemic disease in another group, cross-continental trade and travel resulted in intense epidemics as globalization intensified with extensive cross-continental trading in the 1500s (McNeill 1976; Zinsser 1935). For instance, following its introduction, the Black Death eliminated at least a quarter of the European population (approximately 25 million people) in the 1300s (Laird 1989).

The period of urban development can also be characterized by the exploration and expansion of populations into new areas, which resulted in the introduction of novel diseases to groups that had little resistance to them (McNeill 1976). For example, European air-borne smallpox and measles destroyed millions of Native Americans following contact (Dobyns 1983; Ramenofsky 1987, 1993). The exchange of disease can be a two-way street, however. The exploration of the New World may have been the source of the treponemal infection that was transmitted to the Old World and resulted in the first European epidemic of venereal syphilis (Baker and Armelagos 1988). Treponemal infection in the New World may have been endemic and not sexually transmitted (Rothschild et al. 2000)—that is, the disease had a different mode of transmission when introduced into the Old World. In the Old World, sexual transmission of the treponeme created a different environment for the pathogen, which resulted in a more severe and acute infection. Furthermore, crowding in the urban centers created changes in sexual practices such as prostitution, and an increase in sexual promiscuity may have been a factor in the new venereal transmission of the pathogen (Hudson 1965). Claims that pre-Columbian syphilis existed in Europe have been made in response to the claims of New World origin of the disease (Dutour et al. 1994). The resolution of this debate may await the recovery of material that can be identified as treponemal pathogen from Old World, pre-Columbian archaeological bone.

The process of industrialization, which began a little over two hundred years ago, led to an even greater environmental and social transformation. By 1800, London had one million inhabitants, and other huge city centers soon developed. City dwellers were forced to contend with industrial wastes and polluted water and air, and urban slums became the focal point for poverty and the spread of disease. Epidemics of smallpox, typhus, typhoid, diphtheria, measles, and yellow fever in urban settings are well documented (Polgar 1964). Tuberculosis and respiratory diseases such as pneumonia and bronchitis are also associated with harsh working situations and crowded living conditions. Urban population centers, with their high mortality, were not able to maintain their population base by the reproductive capacity of those living in cities, and required immigration from rural populations in order to maintain their numbers. Cities have been called the "graveyards of humanity."

First Epidemiological Transition: A Perspective on the Evolution of Social Inequality

Paul Farmer (1996) criticizes what he believes is an undue emphasis on the concept of emerging disease. According to Farmer, emerging diseases are only "discovered" when they have an impact on those who have economic power. Lyme disease (*Borrelia burgdorferi*), according to Farmer, was a problem before suburbanites built homes adjacent to golf courses, creating a habitat agreeable to ticks and affluent humans and drawing media attention. In addition, echoing Meredeth Turshen (1977), Farmer claims that an ecological analysis of emerging disease often focuses on human behavior or microbial changes without considering the political–economic contexts that bring pathogens and humans together. He decries the failure to consider how inequality brings at-risk individuals into disproportionate contact with pathogens. The epidemiological model (Barrett et al. 1998) used in this chapter, however, has always emphasized the importance of political–economic factors.

The relationship between inequality and health can be studied from an evolutionary perspective (Goodman et al. 1995), though the evolution of social inequality has not been studied extensively. Social stratification originally evolved because it brought benefits to emerging elites; intensive agriculture and the production of surplus food allowed some members of society to assume administrative occupations not directly related to the primary means of food production. This dichotomy in labor led to the emergence of classes, with the benefits typically accruing to the wealthy and coming at the expense of the poor (Armelagos and Brown 2002). The gap between classes within and among societies has continued to widen since the Neolithic, accelerating with advances in technology. The disparity within and between nations in the present world order continues, and the prospect that this gap will narrow is unlikely. The gap between those on the top and the bottom of the social hierarchy in the 21st century is greater than ever before in human history (Armelagos and Brown 2002). The analysis of changing patterns of health, disease, and social organization in the prehistoric past allows us to better understand health inequalities in the contemporary world (McGuire and Paynter 1991; Paynter 1989).

When evaluating the disease process, we tend to focus on the pathogens (i.e., microparasites) that use the host as a source of food and energy. While microparasites are the source of disease, there are societal factors that foster pathogen survival to such an extent that they are as effective in promoting disease transmission as the parasites themselves. For example, social stratification is an evolutionary strategy in which one segment exploits other segments of the social system to such a degree that the latter's resources are limited and its health is at risk. McNeil (1976) and Peter J. Brown (1987) describe this process of exploitation as "macroparasitism" and see it as a force in understanding the evolution of disease. The acceleration of technological change since the Neolithic has widened the gap between the rich and the poor, the healthy and the sick, within and between societies.

THE SECOND EPIDEMIOLOGICAL TRANSITION: THE RISE OF CHRONIC AND DEGENERATIVE DISEASE

Traditionally, the term *epidemiological transition* refers to the shift from acute infectious diseases to chronic noninfectious degenerative diseases resulting from an increase in longevity. With increasing developments in technology, medicine, and science, the germ theory of disease causation developed. While there is some controversy regarding the role that medicine played in the decline of some of the infectious diseases (McKeown 1979), a better understanding of the source of infectious disease undoubtedly resulted in increased control over many infectious diseases. The development of immunization curbed many infections and was recently the primary factor in the eradication of smallpox. In the developed nations, a number of other communicable diseases, such as measles, yellow fever, and polio, have diminished in importance. The decrease in infectious diseases and the subsequent reduction in infant mortality have resulted in greater life expectancy at birth. The increase in longevity for adults has resulted in an increase in chronic and degenerative diseases.

In addition, the technological advances that characterize the second epidemiological transition result in an increase in environmental degradation. An interesting characteristic of many of the chronic diseases is that they are particularly prevalent and epidemic-like in transitional societies, or in those populations undergoing the shift from developing to developed modes of production. In developing countries, many of the chronic diseases associated with the epidemiological transition appear first in members of the upper socioeconomic strata (Burkitt 1973) because of their access to Western products and practices.

A unique characteristic of the chronic diseases is their relatively recent appearance in human history as a major cause of morbidity. According to Robert S. Corruccini and Samvit S. Kaul (1983), this is indicative of a strong environmental factor in disease etiology. While biological factors such as genetics are no doubt important in determining who is most likely to succumb to which disease, genetics alone cannot explain the rapid increase in chronic disease. Many diseases of the second epidemiological transition share common, etiological factors related to human adaptation, including diet, activity level, mental stress, behavioral practices, and environmental pollution. For example, the industrialization and commercialization of food often results in malnutrition, especially for those societies in transition from subsistence forms of food provision to agribusiness (see, e.g., Luber, this volume; McElroy, this volume). Many individuals in such societies abandon subsistence farming but do not have the economic capacity to purchase food that meets their nutritional requirements (Fleuret and Fleuret 1980). Obesity, linked to the increasing incidence of heart disease and diabetes, is considered to be a common form of malnutrition in developed countries and is a direct result of an increasingly sedentary lifestyle in conjunction with steady or increasing caloric intakes.

Recently, much attention has been focused on the detrimental effects of industrialization on the international environment, including water, land, and atmosphere. Massive industrial production of commodities has caused pollution. There is an increasing concern over the health implications of contaminated water supplies, overuse of pesticides in commercialized agriculture, atmospheric chemicals, and the future effects of depleted ozone on human health and food production. At no other time in human history have the changes in the environment been more rapid and so extreme. Increasing incidence of cancer among young people and the increase in respiratory disease have been implicated in these environmental changes. Problems associated with urbanization and industrialization will become ever more important as the concentration of urban populations grows. In a 2001 briefing paper, the

World Health Organization (2001) reports that 47% of the world's population (2.9 billion people) in 2000 lived in an urban setting. In 30 years, that number will increase to 60%.

EMERGING DISEASE IN THE THIRD EPIDEMIOLOGICAL TRANSITION

Human populations are in the midst of the third epidemiological transition, characterized by both the emergence of new diseases and the reemergence of newly antibiotic-resistant familiar infections on a global scale. This contemporary transition does not preclude infections typical of the first epidemiological transition, resulting in the presence of a number of infectious diseases both old and new; the World Health Organization (1995) reports that of the 50 million deaths each year, 17.5 million are the result of infectious and parasitic disease. The organization also states that two billion people in the world are infected with hepatitis B virus, and two billion of the world's population have tuberculosis (to which eight million cases and three million deaths are attributed every year). In the last thirty years, 40 million people have become infected with HIV, of whom 3 million have died. Many infectious diseases, including those just mentioned, are easily carried over long distances by travelers, and mobility is constantly increasing. In the early 1990s, more than 500 million people annually crossed international borders, and an estimated 20 million refugees and 30 million displaced persons existed worldwide (Wilson 1995); studies have shown that the mobility of the average Westerner has increased dramatically in the last two hundred years (Grubler and Nakicenovic 1991; Guest and Jones, this volume).

D. Satcher (1995) and J. Lederberg (1997) list almost 29 diseases that have emerged in the last 28 years. A list of the most recent emerging diseases include rotavirus (1973), parvovirus B19 (1975), *Cryptosporidium parvum* (1976), Ebola virus (1977), *Legionella pneumophila* (1977), Hantavirus (1977), *Campylobacter jejuni* (1977), HTLV I (1980), *Staphylococcus aureus* toxin (1981), *Escherichia coli* 0157:h7 (1982), HTLV II (1982), *Borrelia burgdorferi* (1982), HIV (1983), *Helicobacter pylori* (1983), *Enterocytozoon bieneusi* (1985) human herpesvirus-6 (1988), hepatitis E (1988) *Ehrlichia chafeensis* (1989), hepatitis C (1989), Guanarito virus (1991), *Encephalitozoon hellen* (1991), new species of *Babesia* (1991), *Vibrio cholerae* 0139 (1992), *Bartonella (Rochalimaea) henselea* (1992), Sin Nombre virus (1993), *Encephalitozoon cuniculi* (1993), Sabia virus (1994), and HHV-8 (1995). Although it may be argued that the long list of pathogens described for the first time in recent years reflects our growing ability to recognize and characterize the agents of disease, the number of novel infections is undoubtedly rising as new technological and trading practices expose humans and pathogens to each other for the first time. For instance, the first outbreak of Nipah virus, in 1998–1999, resulted in 257 cases of acute illness and 100 deaths in Malaysia. This virus was not able to infect humans until pig farming intensified, becoming a major source of income for the country. The novel combination of bats, fruit trees, pig farms, and human labor allowed transmission of the bat-borne virus, first into pigs, as they ingested guano that fell from bats roosting in fruit trees surrounding their pens, and then into humans, as pigs transmitted the disease to their human caregivers. Although bats have no doubt carried Nipah virus for quite some time, the pathogen was never a threat to humans until anthropogenic changes brought them in close proximity to it for the first time (Chua 2003).

The Institute of Medicine (Lederberg et al. 1992) reports that the recent emergence of infectious disease is the result of an interaction of social, demographic, and environmental changes in a global ecology and in the adaptation and genetics of the microbe. Similarly, S. S. Morse (1995) sees emerging disease as a result of demographic changes, international commerce and travel, technological changes, the breakdown of public health measures, and microbial adaptation. Among the ecological changes that Morse describes are the agricultural development projects, dams, deforestation, floods, droughts, and climatic changes that have resulted in the emergence of diseases such as Argentine hemorrhagic fever, Korean hemorrhagic fever (Hantaan), and Hantavirus pulmonary syndrome. Human demographic behavior has been a factor in the spread of dengue, as well as HIV and other sexually transmitted diseases.

Though the long list of exotic newly emerged diseases is impressive, more familiar infections newly recast in antimicrobial-resistant roles may prove to have a greater impact on our lives. Antimicrobial-resistant pathogens have made efficient use of global networks in their spread, and some of the infections they cause are virtually untreatable. In one dramatic example recently reported, multiple-drug resistant pneumococci were brought to Iceland from children returning from vacation in Spain. These pneumococci became established in child care centers in Iceland, and subsequently the penicillin-resistance rate on the island of such infections increased from 1% in 1988 to

17% in 1993 (Livermore 2003). As travel and contact between countries increase, so does the spread of antimicrobial-resistant pathogens.

The factors that govern whether a newly emerged resistant strain will spread or disappear, and whether a particular bacterium is prone to developing resistance, remain poorly understood. Many outbreaks of multiple antibiotic-resistant strains seem to be limited to a small area or set of hosts: David M. Livermore (2003) argues that most outbreaks involve only a few patients in a single hospital unit, usually where the most vulnerable (and heavily medicated) patients are gathered. Other outbreaks, however, affect significantly more people. *Stapholococcus aureus* is a normally harmless bacterium capable of causing a variety of severe problems in those with compromised immune systems; infection can result in toxic shock syndrome, food poisoning, and bacteremia among other disorders. In the United Kingdom, methicillin-resistant *S. aureus* held steady at 1% to 3% during the years 1989–1993; however, concurrent with the emergence of two epidemic strains, resistance rapidly increased to 42% by 2000 (Livermore 2003). In addition, limitation of a resistant strain to an intensive care unit is unlikely in developing nations, where the most important pathogens are community acquired. In sub-Saharan African countries, *Shigella dysentariae*, the pathogen often responsible for epidemic dysentery, has become resistant to all common antibiotics except the fluoroquinolones, which are 400% more expensive than the previous drug used to treat the infection (Howard 2003).

One method by which antibiotic-resistant genes spread is via the plasmid, an extrachromosomal piece of DNA that can be passed to other bacteria either vertically (from original cell to progeny cells after reproduction) or horizontally (from one bacteria to another nearby). Plasmids are able to accrue many genes for resistance to various antimicrobials and to transfer these same genes simultaneously to a more permanent home on the chromosome. They can be transferred from bacterium to bacterium inside the gut of a human or animal host, or anywhere that bacteria come into close contact with one another. Plasmids may be transferred between completely different types of bacteria (Livermore 2003) and even between different species of host organism (Scott 2002). Features such as these make plasmid transfer a particularly effective way of disseminating multidrug resistance among various strains of bacteria and are at the root of the problems that resistance poses to health care today.

The role of humans in the development of antibiotic resistance by way of medical and agricultural practices is well established. Humans are clearly "the world's greatest evolutionary force" (Palumbi 2001). Multinational organizations such as the World Health Organization face a challenge in providing medication to all in need of it and ensuring that medications are used effectively and responsibly to curb the emergence of dangerous and costly resistant strains. The World Bank is the biggest funder for pharmaceuticals worldwide, but with the majority of funding paying for the actual medicine, very little money is allocated to educating local health workers in the proper use of antibiotics (Falkenberg and Tomson 2000).

Antibiotic use to treat human maladies is far from the only contribution to resistant organisms. More than half of the antibiotics produced are used for agricultural purposes. Our acceleration of evolution, according to Stephen R. Palumbi (2001), costs the United States at least $33 billion and as much as $50 billion a year in costs related to antibiotic and pesticide resistance in organisms. Why is combating drug resistance so expensive? As in the aforementioned example of *S. dysenteriae* treatment in Africa, when infections become resistant to the drugs normally used to treat them, more expensive drugs must be resorted to for therapy. Treating multidrug-resistant tuberculosis takes three times as long and is one hundred times more expensive than treatment of a nonresistant case; in addition, the drugs used to treat resistant patients are considerably more toxic, and the levels of resistance to these last-resort treatments are unknown (Pablos-Mendez et al. 2002). Furthermore, alternative drugs are often less effective, despite their higher cost, and the development of new antimicrobials contributes additional cost to the fight against drug-resistant microbes.

CONCLUSION

Because many emerging diseases have their foci in developing and newly affluent countries, a global approach to health care becomes necessary. The globalization of health is not a new concept; international health campaigns began as early as the mid–19th century (Fidler 2001). However, global responses to health threats have received much attention recently. The Centers for Disease Control and Prevention (1994, 1998), following the recommendation of the Institute of Medicine, proposed a plan suggesting a four-pronged attack on emerging disease. First is a need to strengthen infectious surveillance and response in order to better detect and contain infectious agents. Second is the need to research issues raised by these challenges. Third is a need to repair public health infrastructures and increase training. Finally, there is a need to strengthen prevention and control programs "locally, nationally, and globally." The plans of the Centers for Disease Control and Prevention and the Institute of Medicine rely on enhancing the capacity of infrastructure to detect and respond to disease threats. Other projects, such as the "Grand Challenges in Global Health," created by Bill and Melinda Gates and the National Institutes of Health, outline universal health problems and provide funding with the intention of attracting attention from international research teams and benefiting people all over the world (Varmus et al. 2003).

In the inaugural issue of *Emerging Infectious Disease*, David Satcher (1995) is forceful in his assessment of what may be the key in responding to infectious disease. According to Satcher, the role that behavioral sciences have in our efforts to "get ahead of the curve" with respect to emerging infectious disease cannot be overstated. Our knowledge of human behavior must be harnessed to change lifestyle and prevent disease. We understand the cultural practices that allow pathogens to "jump" species barriers and escape their geographic boundaries. The issue of inequality must also be addressed. Extreme poverty, listed near the end of the *International Classification of Disease* and given the code Z59.5, represents the greatest cause of morbidity and mortality in the world (World Health Organization 1995). The World Bank estimates that three billion people in the world live on less than two dollars per day.

The observation that a widening economic and political gap has been the pattern of our post-Neolithic history does not mean that it is inevitable. Some efforts to raise the level of health worldwide have met with great success. Most famously, the eradication of smallpox resulted from an effort that spanned the entire world. Even in living situations characterized by extensive morbidity and mortality, modest outlays can dramatically improve health. For example, each year in the developing world, over 12 million children under the age of five die from causes that could be prevented for a few pennies a day (Armelagos and Brown 2002). In an age when infectious diseases easily traverse the paths between countries, it becomes increasingly important to meet the challenges raised by global health efforts.

NOTE

Kristin Harper's work is supported by a predoctoral fellowship from the Howard Hughes Medical Institute.

REFERENCES

Armelagos, George J.
1990   Health and Disease in Prehistoric Populations in Transition. *In* Disease in
        Populations in Transition. A. C. Swedlund and G. A. Armelagos, eds.
        Pp. 127–144. New York: Bergin and Garvey.
1991   Human Evolution and the Evolution of Human Disease. Ethnicity and
        Disease 1(1):21–26.
1997   Disease, Darwin and Medicine in the Third Epidemiological Transition.
        Evolutionary Anthropology 5(6):212–220.
1998   The Viral Superhighway. The Sciences 38(1):24–30.

Armelagos, George J., and K. Barnes
1999   The Evolution of Human Disease and the Rise of Allergy: Epidemiological
        Transitions. Medical Anthropology 18(2):187–213.

Armelagos, George J., Kathleen C. Barnes, and James Lin
1996   Disease in Human Evolution: The Re-Emergence of Infectious Disease in
        the Third Epidemiological Transition. AnthroNotes 18(3):1–7.

Armelagos, George J., and Peter J. Brown
2002   The Body as Evidence; the Body of Evidence. *In* The Backbone of History.
        R. Steckels and J. Rose, eds. Pp. 593–602. Cambridge: Cambridge University
        Press.

Audy, J. R.
1961   The Ecology of Scrub Typhus. *In* Studies in Disease Ecology. J. M. May, ed.
        Pp. 389–432. Studies in Medical Geography, 2. New York: Hafner.

Baker, Brenda, and George J. Armelagos
1988   Origin and Antiquity of Syphilis: A Dilemma in Paleopathological
        Diagnosis and Interpretation. Current Anthropology 29(5):703–737.

Barnes, Kathleen C., George J. Armelagos, and Steven C. Morreale
1999   Darwinian Medicine and the Emergence of Allergy. *In* Evolutionary
        Medicine. W. Trevethan, J. McKenna, and E. O. Smith, eds. Pp. 209–244.
        New York: Oxford University Press.

Barrett, Ronald, Christopher W. Kuzawa, Thomas McDade, and George J. Armelagos
1998   Emerging Infectious Disease and the Third Epidemiological Transition.
        In Annual Review Anthropology, 27. W. Durham, ed. Pp. 247–271. Palo
        Alto, CA: Annual Reviews.

Black, Francis L., W. J. Hierholzer, F. Pinheiro, A. S. Evans, J. P. Woodall, E. M. Opton, J. E. Emmons, B. S. West, G. Edsall, W. G. Downs, and G. D. Wallace
1974   Evidence for Persistence of Infectious Agents in Isolated Human Populations. American Journal of Epidemiology 100:230–250.

Brown, Peter J.
1987   Microparasites and Macroparasites. Cultural Anthropology 2:155–171.

Burkitt, D. P.
1973   Some Disease Characteristics of Modern Western Medicine. British Medical Journal 1:274–278.

Burnet, F. M.
1962   Natural History of Infectious Disease. Cambridge: Cambridge University Press.

Centers for Disease Control and Prevention
1994   Addressing Emerging Infectious Disease Threats: A Prevention Strategy for the United States. Atlanta: U.S. Public Health Service.
1998   Preventing Emerging Infectious Diseases: A Strategy for the 21st Century. Atlanta: U.S. Public Health Service.

Chandler, Tertius
1987   Four Thousand Years of Urban Growth: An Historical Census. Lewiston, NY: Edward Mellen Press.

Chua, K. B.
2003   Nipah Virus Outbreak in Malaysia. Journal of Clinical Virology 26(3):265–275.

Cockburn, T. A.
1967a   The Evolution of Human Infectious Diseases. In Infectious Diseases: Their Evolution and Eradication. T. A. Cockburn, ed. Pp. 84–107. Springfield, IL: Charles C. Thomas.
1967b   Infections of the Order Primates. In Infectious Diseases: Their Evolution and Eradication. T. A. Cockburn, ed. Pp. 38–49. Springfield, IL: Charles C. Thomas.
1971   Infectious Disease in Ancient Populations. Current Anthropology 12(1):45–62.

Cohen, M. N., and G. J. Armelagos, eds.
1984   Paleopathology at the Origins of Agriculture. New York: Academic Press.

Corruccini, Robert. S., and Samvit S. Kaul
1983   The Epidemiological Transition and the Anthropology of Minor Chronic Non-Infectious Diseases. Medical Anthropology 7:36–50.

Desowitz, R. S.
1980   Epidemiological-Ecological Interactions in Savanna Environments. *In* Human Ecology in Savanna Environments. D. R. Harris, ed. Pp. 457–477. London: Academic Press.

Dicke, B. H.
1932   The Tseste Fly's Influence on South African History. South African Journal of Science 29:792.

Dobyns, Henry
1983   Their Numbers Become Thinned: Native American Population Dynamics in Eastern United States. Knoxville: University of Tennessee Press.

Dutour, O., Gy. Pálfi, J. Bérato, and J.-P. Brun
1994   L'origine de la Syphilis en Europe—Avant ou Aprés 1493? Paris: Errance.

Falkenberg, T., and G. Tomson
2000   The World Bank and Pharmaceuticals. Health Policy and Planning 15(1):52–58.

Farmer, Paul
1996   Social Inequalities and Emerging Infectious Diseases. Emerging Infectious Diseases 2(4):259–269.

Fidler, David P.
2001   The Globalization of Public Health: The First 100 Years of International Health Diplomacy. Bulletin of the World Health Organization 79(9):842–849.

Fleuret, P., and A. Fleuret
1980   Nutrition, Consumption and Agricultural Change. Human Organization 39(3):250–260.

Friedman, Thomas L.
1999   Lexus and the Olive Tree: Understanding Globalization. New York: Farrar, Straus, Giroux.

Georgopapadakou, N. H.
2002   Infectious Disease 2001: Drug Resistance, New Drugs. Drug Resistance Updates 5(5):181–191.

Goldstein, G.
2000   Healthy Cities: Overview of a WHO International Program. Reviews on
        Environmental Health 15(1–2):207–214.

Goodman, Alan H., Debra L. Martin, and George J. Armelagos
1995   The Biological Consequences of Inequality in Prehistory. Rivista di
        Anthropologia (Roma) 73:123–131.

Grubler A., and N. Nakicenovic
1991   Long waves, technology diffusion and substitution. Review 14:313–342.

Hasegawa, Hi
1999   Phylogeny, Host-Parasite Relationship and Zoogeography. Korean Journal
        of Parasitology 37(4):197–213.

Hoberg, E. P., N. L. Alkire, A. de Queiroz, and A. Jones
2001   Out of Africa: Origins of the Taenia Tapeworms in Humans. Proceedings
        of the Royal Society of London–Series B: Biological Sciences
        268(1469):781–787.

Hoberg, E. P., A. Jones, R. L. Rausch, K. S. Eom, and S. L. Gardner
2000   A Phylogenetic Hypothesis for Species of the Genus Taenia (Eucestoda:
        Taeniidae). Journal of Parasitology 86(1):89–98.

Howard, David H., R. Douglas Scott II, Randall Packard, and DeAnn Jones
2003   The Global Impact of Drug Resistance. Clinical Infectious Diseases
        36(Suppl. 1):S4–S10.

Hudson, E. H.
1965   Treponematosis and Man's Social Evolution. American Anthropologist
        67:885–901.

Hughes, James M.
2001   Emerging Infectious Diseases: A CDC Perspective. Emerging Infectious
        Diseases 7(3):494–496.

Johansson, S. R.
1991   The Health Transition: The Cultural Inflation of Morbidity during the
        Decline of Mortality. Health Transition Review 1:39–68.
1992   Measuring the Cultural Inflation of Morbidity during the Decline in
        Mortality. Health Transition Review 2(1):78–89.

Joy, Deirdre A., Xiaorong Feng, Jianbing Mu, Tetsuya Furuya, Kesinee Chotivanich,
        Antoniana U. Krettli, May Ho, Alex Wang, Nicholas J. White, Edward Suh, Peter
        Beerli, and Xin-zhuan Su

2003   Early Origin and Recent Expansion of Plasmodium Falciparum. Science 300:318–321.

Kenzer, M.
2000   Healthy Cities: A Guide to the Literature. Public Health Reports 115(2–3):279–289.

Kittler, R., M. Kayser, and M. Stoneking
2003   Molecular Evolution of Pediculus Humanus and the Origin of Clothing. Current Biology 13:1414–1417.

Kombe, G. C., and D. M. Darrow
2001   Revisiting Emerging Infectious Diseases: The Unfinished Agenda. Journal of Community Health 26(2):113–122.

Kunitz, S. J.
1991   The Personal Physician and the Decline of Mortality. *In* The Decline of Mortality in Europe. D. R. Sclofield and A. Bideau, eds. Pp. 248–262. Oxford: Clarendon Press.

Laird, M.
1989   Vector-Borne Disease Introduced into New Areas due to Human Movements: A Historical Perspective. *In* Demography and Vector-Borne Diseases. M. W. Service, ed. Pp. 17–33. Boca Raton, FL: CRC Press.

Lambrecht, F. L.
1964   Aspects of Evolution and Ecology of Tsetse Flies and Trypanosomiasis in Prehistoric African Environments. Journal of African History 5:1–24.
1967   Trypanosomiasis in Prehistoric and Later Human Populations: A Tentative Reconstruction. *In* Diseases in Antiquity. D. Brothwell and A. T. Sandison, eds. Pp. 132–151. Springfield, IL: Charles C. Thomas.
1980   Paleoecology of Tsetse Flies and Sleeping Sickness in Africa. Proceedings of the American Philosophical Society 124(5):367–385.
1985   Trypanosomes and Hominid Evolution. Bioscience 35(10):640–646.

Lederberg, J.
1997   Infectious Disease as an Evolutionary Paradigm. Emerging Infectious Diseases 3(4):417–423.

Lederberg, Joshua, Robert E. Shope, and Stanley C. Oaks, eds.
1992   Emerging Infections: Microbial Threats to Health in the United States. Washington, DC: National Academy Press.

Livermore, David M.
2003   Bacterial Resistance: Origins, Epidemiology, and Impact. Clinical Infectious
         Diseases 36(Suppl. 1):S11–S23.

Livingstone, F. B.
1958   Anthropological Implications of Sickle-Cell Distribution in West Africa.
         American Anthropologist 60:533–562.
1971   Malaria and Human Polymorphisms. Annual Review of Genetics 5:33–64.
1983   The Malaria Hypothesis. In Distribution and Evolution of Hemoglobin and
         Globin Loci, 4. J. Bowman, ed. Pp. 15–44. New York: Elsevier.

Mah, M. W., and Z. A. Memish
2000   Antibiotic Resistance. An Impending Crisis. Saudi Medical Journal
         21(12):1125–1129.

McGuire, Randall, and Robert Paynter, eds.
1991   The Archaeology of Inequality. Oxford: Basil Blackwood.

McKeown, T.
1976   The Modern Rise of Population. New York: Academic Press.
1979   The Role of Medicine: Dream, Mirage or Nemesis. Princeton, NJ: Princeton
         University Press.

McMichael, A. J.
2000   The Urban Environment and Health in a World of Increasing
         Globalization: Issues for Developing Countries. Bulletin of the World
         Health Organization 78(9):1117–1126.

McNeill, W. H.
1976   Plagues and People. Garden City, NY: Anchor/Doubleday.
1978   Disease in History. Social Science and Medicine 12:79–81.

Morse, S. S.
1995   Factors in the Emergence of Infectious Diseases. Emerging Infectious
         Diseases 1:7–15.

Omran, A. R.
1971   The Epidemiologic Transition: A Theory of the Epidemiology of
         Population Change. Millbank Memorial Fund Quarterly 49(4):509–538.
1975   The Epidemiologic Transition in North Carolina during the Last 50 to 90
         Years: II. Changing Patterns of Disease and Causes of Death. North
         Carolina Medical Journal 36(2):83–88.

1977   A Century of Epidemiologic Transition in the United States. Preventive
        Medicine 6(1):30–51.
1982   Epidemiological Transition. *In* International Encyclopedia of Population.
        J. A. Ross, ed. Pp. 172–183. London: Free Press.
1983   The Epidemiologic Transition Theory: A Preliminary Update. Journal of
        Tropical Pediatrics 29(6):305–316.

Pablos-Mendez, A., D. K. Gowda, and T. R. Frieden
2002   Controlling Multi-Drug Resistant Tuberculosis and Access to Expensive
        Drugs: A Rational Framework. Bulletin of the World Health Organization
        80(6):489–495.

Palumbi, Stephen R.
2001   Humans as the World's Greatest Evolutionary Force. Science
        293(5536):1786–1790.

Paynter, Robert
1989   The Archaeology of Equality. Annual Review of Anthropology 18:369–399.

Polgar, Steven
1964   Evolution and the Ills of Mankind. *In* Horizons of Anthropology. S. Tax, ed.
        Pp. 200–211. Chicago: Aldine.

Ramenofsky, Ann
1987   Vectors of Death: The Archaeology of European Contact. Albuquerque:
        University of New Mexico Press / Center for Documentary Studies at Duke
        University.
1993   Diseases in the Americas, 1492–1700. *In* The Cambridge World History of
        Human Disease. K. Kiple, ed. Pp. 417–427. New York: Cambridge
        University Press.

Robertson, Roland
1992   Globalization: Social Theory and Global Culture. Newbury Park, CA: Sage.

Rothschild, B. M., F. L. Calderon, A. Coppa, and C. Rothschild
2000   First European Exposure to Syphilis: The Dominican Republic at the Time
        of Columbian Contact. Clinical Infectious Diseases 31(4):936–941.

Satcher, D.
1995   Emerging Infections: Getting Ahead of the Curve. Emerging Infectious
        Diseases 1(1):1–6.

Schofield, R., and D. Reher
1991   The Decline of Mortality in Europe. *In* The Decline of Mortality in Europe.
        R. Schofield, D. Reher, and A. Bideau, eds. Pp. 1–17. Oxford: Clarendon
        Press.

Scott, K. P.
2002   The Role of Conjugative Transposons in Spreading Antibiotic Resistance
        between Bacteria That Inhabit the Gastrointestinal Tract. Cell and
        Molecular Life Sciences 59:2071–2082.

Shipman, Pat
2002   A Worm's View of Human Evolution. American Scientist 90(6):508–510.

Slifko, T. R., H. V. Smith, and J. B. Rose
2000   Emerging Parasite Zoonoses Associated with Water and Food. International
        Journal for Parasitology 30(12–13):1379–1393.

Sprent, J. F. A.
1962   Parasitism, Immunity and Evolution. *In* The Evolution of Living
        Organisms. G. S. Leeper, ed. Pp. 149–165. Melbourne: Melbourne
        University Press.
1969a   Evolutionary Aspects of Immunity of Zooparasitic Infections. *In* Immunity
        to Parasitic Animals, vol. 1. G. J. Jackson, ed. Pp. 3–64. New York: Appleton.
1969b   Helminth "Zoonoses": An Analysis. Helminthology Abstracts 38:333–351.

Steckels, Richard, and Jerome Rose, eds.
2002   The Backbone of History. Cambridge: Cambridge University Press.

Stevens, J., H. Noyes, and W. Gibson
1998   The Evolution of Trypanosomes Infecting Humans and Primates.
        Memorias do Instituto Oswaldo Cruz 93(5):669–676.

Svanborg-Eden, Catrina, and Bruce R. Levin
1990   Infectious Disease and Natural Selection in Human Populations. *In* Disease
        in Populations in Transition. A. C. Swedlund and G. J. Armelagos, eds.
        Pp. 31–48. New York: Bergin and Garvey.

Tishkoff, Sarah A., Robert Varkonyi, Nelie Cahinhinan, Salem Abbes, George
   Argyropoulos, Giovanni Destro-Bisol, Anthi Drousiotou, Bruce Dangerfield,
   Gerard Lefranc, Jacques Loiselet, Anna Piro, Mark Stoneking, Antonio Tagarelli,
   Giuseppe Tagarelli, Elias H. Touma, Scott M. Williams, and Andrew G. Clark
2001   Haplotype Diversity and Linkage Disequilibrium at Human G6PD: Recent
        Origin of Alleles That Confer Malarial Resistance. Science
        293(5529):455–462.

Tsouros, A. D.
2000   Why Urban Health Cannot Be Ignored: The Way Forward. Reviews on Environmental Health 15(1–2):267–271.

Turshen, Meredeth
1977   The Political Ecology of Disease. Review of Radical Political Economics 9(1):45–60.

Varmus, H., R. Klausner, E. Zerhouni, T. Acharya, A. S. Daar, and P. A. Singer
2003   Grand Challenges in Global Health. Science 302(5644):398–399.

Waters, W. F.
2001   Globalization, Socioeconomic Restructuring, and Community Health. Journal of Community Health 26(2):79–92.

Weber, D. J., and W. A. Rutala
1999   Zoonotic Infections. Occupational Medicine 14(2):247–284.

Wilson, Mary E.
1995   Travel and the Emergence of Infectious Diseases. Emerging Infectious Diseases 1(2):39–46.

Woods, R.
1990   The Role of Public Health in the Nineteenth-Century Mortality Decline. *In* What We Know about Health Transition: The Cultural, Social, and Behavioral Determinants of Health. J. Caldwell, S. Findley, P. Caldwell, G. Santow, W. Cosford, J. Braid, and D. Broers-Freeman, eds. Pp. 110–115. Proceedings of an International Workshop, 1. Canberra: Health Transition Centre.

World Health Organization
1995   Executive Summary. The World Health Report: Bridging the Gaps. Geneva: World Health Organization.
2001   The World Health Report 2001. Geneva: World Health Organization.

Zimmer, C.
2001   Genetic Trees Reveal Disease Origins. Science 292(5519):1090–1093.

Zinsser, Hans
1935   Rats, Lice and History. Boston: Little, Brown and Company.

3

Poverty and Violence, Hunger and Health: A Political Ecology of Armed Conflict

Thomas L. Leatherman

Worldwide patterns of health are intricately linked to changing physical, so-cial, and economic environments in contexts of global capitalism. While this has been the case for at least the past five hundred years, the linkages have drawn tighter in recent decades. Global capitalism and growing social and economic inequalities have amplified the conditions for environmental degra-dation, impeded access to resources, food insecurity, social conflicts, and pop-ulation displacement—with serious ramifications for human nutrition and health (Farmer 1999, 2003; Garett 1994, 2000; Kim et al. 2000). Poor countries and peoples are most vulnerable to disparities in nutrition and health, but among wealthy nations as well, social, economic, and political inequalities can be reflected in health disparities (Sen 1993, 1999; Marmot et al. 1997).

Given the vast improvements in health and health care over the past half-century and the unprecedented growth of the global economy, the alarmingly high numbers of the poor, hungry, starving, and sick worldwide point to struc-tured inequities at global and local scales. Indeed, the links between poverty, in-equalities, and health have become central features in even mainstream analyses of conditions of vulnerability in a global economy (Feacham 2000; Food and Agriculture Organization [FAO] 2002; World Health Organization [WHO] 2002). What is also different in recent years is an expanded view of the relevant causes and contexts of vulnerability. Increasingly, we find that issues of envi-ronmental degradation and pollution, armed conflicts, violence, population displacement, food insecurity, and chronic and infectious disease are tightly woven into processes of globalization and, in synergy, constitute sources and indicators of vulnerability in human populations (Homer-Dixon 1994; Messer et al. 1998; WHO 2002).

In this chapter, I have two objectives. One is to outline interactions of environment and health in contexts of globalization and to suggest a political–ecological perspective as one avenue for addressing these interactions. The second objective is to illustrate how political ecology helps frame analyses of complex interactions among inequalities, environmental scarcities, food insecurity, hunger, and health. The context for demonstrating these interactions is the all-too-common example of armed conflicts, drawing on the work of Ellen Messer and others on food insecurity and conflicts (Messer et al. 1998; FAO 2002; Cohen and Pinstrup-Andersen 1999; Teodosijevic 2003). I then turn to the context and consequences of the Sendero Luminoso (Shining Path) revolt in Peru (see, e.g., McClintock 1984, 1989; Poole and Rénique 1992; Stern 1998). I link these findings to my research on health and household economy in the southern Andean highlands (Leatherman 1994, 1996, 1998) in order to provide local and ethnographically specific insights into the conditions of poverty and inequality that underlie conflicts.

GLOBALIZATION, INEQUALITY, ENVIRONMENT, AND HEALTH

It has become commonplace in recent years to recognize the extent to which everything is global—from markets, factories, and commodities (and their producers and consumers); to transnational corporations, trade agreements (e.g., General Agreement on Tariffs and Trade, North American Free Trade Agreement), and lending institutions; to systems of communication and transportation; to health, disease, and food insecurity (Blim 1992; Robbins 2002). As I. Wallerstein (1974) has argued, we have lived in an interconnected world system since the 16th century, incorporating multiple regions, cultures, and people into a global capitalist market characterized by a worldwide division of labor. Although globalization is not new, the sheer breadth and depth of these interconnections have intensified in the past three decades.

While global markets and institutions are seen by many as an avenue toward improved economic and health conditions, others point to a rise in structural inequalities, poverty, environmental scarcity, and poor health, all in contexts of global capitalism. M. L. Blim (1992), for example, links globalization to patterns of uneven development, debt crisis, austerity packages requiring diminished social programs, heightened inequalities, and the creation of a context and space for popular protest. Transnational corporations, for instance, make up over half of the one hundred largest international economies (Gersham and Irwin 2000; Robbins 2002), and according to the United Nations Development Programme (Latham and Beaudry 2001) the wealth of the three richest people in the world in 1998 was more than the combined gross national product of the 48 least developed countries. In 1997, the income gap between the fifth of the world's population living in the wealthiest nations and the fifth living in poorest nations was 74 to 1, up from 30 to 1 in 1960 (Latham and Beaudry 2001).

Social and economic inequalities translate into health and nutritional inequalities. Despite remarkable achievements in global health over the last four decades (e.g., an 18-year increase in average life expectancy and a 60% decrease in infant mortality), there remains a 16-fold difference in infant mortality between the 26 wealthiest nations and the 48 least developed countries. Half of the people in the world's poorest 46 nations are without access to modern health care; three billion do not have access to sanitation facilities; one billion do not have access to safe drinking water; and 600 million live in what the World Health Organization has termed "life-threatening" homes and neighborhoods in urban environments (Millen et al. 2000). Amayrta Sen (1993) has noted that in the world's wealthiest nation, the United States, African Americans have a lower chance of reaching advanced ages than do counterparts born in the immensely poorer economies of China or the Indian state of Kerala. The causal influences, he notes, include relative inequities in wealth but also in medical coverage, public health care, school education, law and order, and the prevalence of violence.

While equitable access to modern health care is clearly a problem with such disparities, problems of food security and basic nutrition are another critical link between poverty, inequality, and poor health. Nutritionists Michael Latham and Micheline Beaudry note that while "globalization has been offered as the answer to improving the economies of poor countries . . . current evidence shows that globalization as it has been practiced has been accompanied by increases in inequity and a reduction of poor nations' ability to achieve national or local food and nutrition security" (2001:599). Efforts at reducing malnutrition that focus primarily on increasing production or technological solutions (e.g., micronutrient fortification) rather than on improving access to resources, food, and health care have done little to alleviate the problem. Indeed, they note that the adequacy of the food supply has increased in every region except Africa but that food intake has not increased for the poor. Hence, there are 840 million undernourished people in the world, 95% of whom live in developing countries (FAO 2002). Six million children under the age of five die each year from hunger. Relatively few die in famines, under the gaze of the world media. Rather, the majority die from chronic malnutrition that depletes both energy and immune system—a "covert famine" (FAO 2002) or, stated more broadly, a "silent violence" (Watts 1983).

Environmental degradation and unequal and unjust access to resources and entitlements are considered to be among the most important factors shaping global health and are important links between inequality and health (see, e.g., Donohoe 2003). Moreover, the characteristics of a modern global economy, with rapid and extensive means of communication and transportation, increasingly draw tighter the linkages between markets, commodities, environments, people, and pathogens (see, e.g., Eyles and Consitt, this volume). What were once relatively self-contained, unconnected, unpolluted ecosystems are now part of an interconnected world ecology. In this vein, Laurie Garrett (1994, 2000) has convincingly argued that the health of every nation in the world is linked to every other as a basic epidemiological fact of a new global world.

Ecological models of disease direct us to examine the relationship between humans, environment, and disease. But what has become clearer is the degree that political, economic, and social factors shape local environments and, in turn, the relationship between humans, environment, and health. The breadth and depth of these interconnections suggest the need for perspectives that focus on human–environment interactions, hunger, disease, and malnutrition, in terms of power relations that shape poverty and inequality, as well as political, socioeconomic, and environmental marginalization.

Political Ecology of Health

To link local-level vulnerabilities to structures of poverty and inequality played out in human–environment interactions, anthropologists, geographers, and others have turned to ways to integrate ecology and political economy into a "political ecology" approach (Peet and Watts 1996; Bryant and Bailey 1997 for partial reviews). Political ecology explicitly connects issues of power and inequality (from political–economy) with human–environment interactions (the concern of ecological anthropology) and addresses these relationships at the intersection of the global and the local. Components of a political ecology approach that frame its orientation and distinguish it from other economic and ecological approaches include a focus on broad contexts, historical depth, and the interaction of structure and agency in addressing human–environment relationships (see Kalipeni and Oppong 1998; Bryant and Bailey 1997; Leatherman and Thomas 2001). Thus, society and environment relationships are placed in the context of local histories and ecologies, and these are in turn shaped by broader historical patterns ranging from how local populations transform local ecologies in production to how global forces shape these processes.

Political ecology perspectives have been used first and perhaps foremost in studies of land use and the politicization of environmental problems (e.g., Blaike and Brookfield 1987), but they cast a much broader net into studies of health, food security and nutrition, environmental activism, environmental justice movements, and discourses of development (Peet and Watts 1996). A political ecology of health approach frames the study of health in terms of global–local connections, capitalist relations, resource scarcities, inequalities and poverty, and human rights, as well as disease vectors, hygiene, pathogenic exposure, and hunger and famine. It directs us to examine the political–economic history (often of colonialism) and subsequent global–local contexts that structure regional- and local-level inequalities, injustice, and crisis. In the context of this chapter, it directs us to examine the human–environment dimensions of crises—environmental struggles, subsistence crises, the interplay of poverty, hunger and health—and in what situations these realities contribute to the emergence of social unrest and conflict. Finally, while the causes of conflict are deep and structural (i.e., structured inequalities, poverty, hunger, illness, and violence), a political ecology approach provides greater room for human agency in understanding when collective revolt and violence to change systems, not accommodate them, become the logic for action.

POLITICAL ECOLOGY OF ARMED CONFLICT

Conflict is one of the most common causes of food insecurity. The displacement of people and disruption of agricultural production and food distribution leave tens of millions of people at risk of hunger and famine. (p. 22)

The reduced productivity, truncated working lives and suffocated opportunities of 799 million hungry people in the developing world hamstring economic progress and fuel environmental degradation and conflict at the national and international levels. (p. 4)

—Food and Agriculture Organization, *The State of Food Insecurity in the World, 2002*

Tensions ripen into violent conflict especially where economic conditions deteriorate and people face subsistence crises. Hunger causes conflict when people feel they have nothing more to lose and so are willing to fight for resources, political power, and cultural respect. (p. 383–384)

—M. J. Cohen and P. Pinstrup-Andersen, *"Food Security and Conflict"*

As these statements indicate, food insecurity and hunger not only are symptoms of impoverished and marginalized peoples and environments but serve as a catalyst of marginalization and vulnerability. One aspect of this vulnerability is the human and environmental toll of armed conflicts that have become all too commonplace in the past three decades. At least 47 countries in Africa, Asia, Latin America, and Europe and the former Soviet Union have experienced armed conflicts since the 1970s (Messer 1998; Stockholm International Peace Research Institute 2002). These conflicts have their roots in histories of inequality and injustice, as well as the synergism of poverty, hunger, and illness—all of which can be exacerbated in contexts of globalization. Conflict takes its toll in lost lives, families, communities, property, production, livelihood, environmental degradation, and markets, which in turn can promote food insecurity, hunger, malnutrition, and starvation.

Inequality and Insecurity as Catalysts for Conflict

The contexts of armed conflict illustrate strong reciprocal effects between poverty and inequality; environmental scarcities and degradation; food insecurity and poor health; and local, national, and international policies and practices that shape many of these processes (see Figure 3.1). Frequently cited causes of conflict include political, social, and economic factors such as

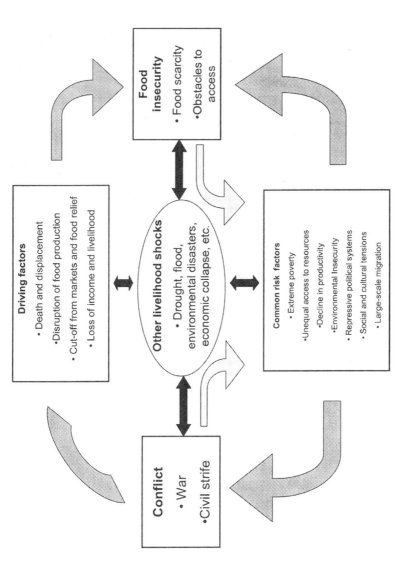

FIGURE 3.1

Interface between conflict and food insecurity (adapted from Food and Agriculture Organization 2002:23).

unequal access to land and control over natural resources, repressive and unjust exercise of power, absence of political representation, and deteriorating social service; racial, ethnic, class, and religious divisions, discrimination, and grouped-based fanaticism; the ready availability of weapons; and rapid demographic change (Messer et al. 1998; Cohen and Pinstrup-Andersen 1999; Teodosijevic 2003; WHO 2002).

Histories of regional armed conflict show deep roots in colonial, postcolonial, contemporary politics, racial–religious exclusion, and socioeconomic discrimination, as well as struggle over strategic resources such as land, water, trade routes, and petroleum (Messer et al. 1998). Also implicated is a discontent over land distribution and a lack of entitlements, precluding access to basic needs and heightening disparities in education, health, food security, and basic human rights. Rapid increases in population and population densities strain all services and, when accompanied by high infant mortality, highlight the absence of basic needs.

T. F. Homer-Dixon (1994) and P. Le Billon (2001) see vulnerability to conflict as being linked to environmental scarcity, control over and competition for natural resources and commodities, and dependency on the export of primary commodities. Homer-Dixon argues that the decrease in quality and quantity of renewable resources, in combination with population growth and unequal access to resources, leads to real and perceived scarcities that lie at the heart of many violent conflicts. Le Billon notes that, especially in the aftermath of the Cold War, natural resources such as oil, gems, and timber provide the revenue for armed conflict and a site of struggle that motivates armed conflict. He cites as examples the struggles to control diamonds in Angola and the Democratic Republic of the Congo, oil in Angola and Sudan, timber and gems in Cambodia and Afghanistan, and coca and the drug trade in Columbia. Vulnerability to conflict is heightened in cases of strong economic dependency on resources as primary export commodities, such as the case of coffee in Rwanda (see Messer et al. 1998:24–25).

Globalization can widen and intensify inequalities and at the same time compromise the role of state, increase privatization, and reduce social safety nets (see, e.g., Guest and Jones, this volume). Such is the case particularly in the widely implemented and criticized structural adjustment programs of the International Monetary Fund and World Bank, which provide much-needed international loans for countries but require agreements to reduce the public sector; open markets; and reduce price subsidies, assistance programs, and other safety nets. Debt and structural adjustment burdens correlate with conflict. Smith (Messer et al. 1998:13) found that 50 of 71 countries receiving structural adjustment loans were experiencing conflict and that 22 of the top 25 developing-country debtors were in conflict, including Argentina, Honduras, Mexico, Peru, the Philippines, and many African countries.

However, while conflicts are linked to real and perceived environmental scarcities, dependence on primary export commodities, and food insecurity in crisis proportions, none of these factors should be seen as being determinative (Gleditsch 1998). Resources are by definition socially constructed. Thus, scarcity is a social and political construction, and the degree to which scarcity leads to conflict usually depends on the degree to which it exacerbates chronic vulnerabilities due to broader inequalities and marginalization. Armed conflict and violence are always underlain by structural violence (Heggenhougen 1995; Farmer 1999, 2003) that robs people of access to resources, creates food insecurities, and strengthens the reciprocal synergism of poverty and poor health. The costs of conflict, in turn, help reproduce and intensify the very conditions that create crisis and desperation and prompt violence as an option for social change.

Costs of Conflict on Environments, Hunger, and Health

Each year for the past 20 years, one million lives were lost to armed conflict (Messer et al. 1998). Most of the victims (perhaps 90%; Stockholm International Peace Research Institute 2002) were civilians, and most of the deaths occurred away from the battlefield, caused by malnutrition and disease and the traumas of conflict and displacement (Messer et al. 1998; Kalipeni and Oppong 1998; Teodosejevic 2003). As recent as 1996, an estimated 80 million people were at risk to hunger and malnutrition due to conflicts and their aftermath. Included in this count were 23 million refugees and an additional 27 million internally displaced peoples (Messer et al. 1998), the majority in Africa and Asia. Displaced populations and refugees are at risk to nutritional deprivation, illness (especially respiratory and gastrointestinal ailments), and violence, in the crowded and unhygienic conditions of emergency camps. Of all refugees and internally displaced peoples, approximately 70% to 80% are women and children, who are particularly vulnerable to psychological trauma, malnutrition and disease, violence, and sexual abuse. Refugees also demand food, water, land, fuel, and other resources that can strain a precarious production system in the receiving country or region to the point of food disaster. Return refugees can often bring disease back to their home regions, the case especially for AIDS in Africa (Kalipeni and Oppong 1998).

Not only do armed conflicts take a toll in lives and property and promote massive population displacement, but they also disrupt food production and food and medicine distribution as well as cause environmental degradation (Cohen and Pinstrup-Andersen 1999; FAO 2002; Messer et al. 1998; Teodosijevic 2003; WHO 2002). War and civil strife were cited as major causes in 15 of the 44 countries that suffered extreme food emergencies in 2001–2002 (FAO 2002). Messer and colleagues (1998) examined measures of food production during war and peace years for 14 African countries and found an average decline in production of 12% during war years—with figures ranging from no declines in Kenya to a 44% reduction in Angola. Similarly, Slobodanka Teodosejivic (2003) estimates a 10% reduction in food production for 38 countries during war years between 1961 and 2000. Such effects extend beyond borders to threaten food security in surrounding regions. For example, hypothetical food-projection estimates through the year 2000 suggested that Mozambique, in absence of conflict, could have made southern Africa self-sufficient in rice (Messer et al. 1998).

The most obvious way in which conflict leads to food insecurity and hunger is through the deliberate use of hunger as a weapon (Messer et al. 1998). Armies starve, kill, and enslave opponents; destroy food supplies, livestock, and other means of production; cut off markets and impede or divert food relief. Moreover, fighting destroys environments and, in the case of landmines, makes them dangerous for years to come (a buried mine can stay active for 50 years). There are currently between 60 million and 119 million mines buried in 70 countries; 10 million of these are in Cambodia, where 1 in every 236 persons is an amputee due to mine-related injury (Messer et al. 1998). Globally, land mines kill or maim two thousand people every month, mostly women and children. An estimated 25 mines per square mile in Rwanda kill and disable farmers, disrupt trade, make cultivation hazardous, hamper resettlement, and thus negatively affect long-term food security (Messer et al. 1998).

In most countries experiencing armed conflict, increased military spending is necessarily paired with a concurrent reduction in expenditures in health, education, food, and other basic needs. Between 1960 and 1994, the developing world imported US$775 billion worth of military supplies. Martin Donohoe notes that

> three hours of world-wide military spending is equal to the [World Health Organization's] annual budget. Three days of US military spending equals that spent on health, education and welfare for children in one year. Three weeks of world arms spending could provide primary health care for all individuals in poor countries including water and hygiene. [2003:580]

Thus, conflict leads to death, human displacement, impaired production, and loss of revenues, all of which accentuate levels of poverty, food insecurity, malnutrition, and disease. Direct impacts on health result from fear and anxiety, as well as hunger and malnutrition, but also from destruction to health care infrastructure and reduced government support of social service, education, and health care (Zwi and Ugalde 1989; Lundgren and Lang 1989; Cervantes et al. 1989).

However, the health impacts of armed conflicts are not straightforward. In El Salvador, despite an all-consuming and often brutal war over a period of 12 years (1980–1992), infant mortality, rates of immunization, and some other common health indicators remained steady or improved slightly (Ugalde et al. 2000). This was in part due to relief efforts from the World Food Program, USAID, a temporary truce between government and opposing forces that allowed immunization campaigns to occur, and the fact that military attacks generally spared health facilities (Ugalde et al. 2000). Yet, while infant mortality declined, neonatal mortality rose during this same period, as did the percentage of those malnourished—the former due to disrupted health service infrastructure and the latter to reduced and disrupted food production and distribution. Moreover, mortality rates among the five hundred thousand *desplazados* (displaced peoples) were 3.4 times greater than those of the general population, and malnutrition and diarrhea were major causes of death for children under five years of age (Lundgren and Lang 1989).

The long-term costs on health due to the trauma and psychosocial stress experienced during war years is still unknown and will emerge only after more study and years of peace. There are indications that levels of suicide and interpersonal violence in postwar societies increase and that various symptoms related to posttraumatic stress disorder are evident in terms of anxiety, sleep disorder, nightmares, depression, and alcoholism (e.g., Cervantes et al. 1989; Hollifield et al. 2002). In El Salvador, frequent problems found among the traumatized in refugee camps included: insomnia, nervous tics, autism, aggressiveness, manic depression, and hyperparanoia; difficulty in learning to read and write (in children); and somatic complaints of frequent headaches, stomach cramps, and colitis (Lundgren and Lang 1989). It is also the case that trauma is experienced and somaticized unevenly and that conventional means of assessing impacts (based on posttraumatic stress scales) may be far from adequate as they are applied in the multiple cultural and social contexts in which conflicts occur (Bracken et al. 1995).

Finally, a critical problem is that food and economic insecurity linger long after fighting has ceased. Conflicts disrupt markets and leave households without sufficient resources to access food, and this leads to food insecurity or poverty-related hunger. Besides basic resource-distribution conflicts, there are also disrupted flows of migrant labor and patterns of remittance. Moreover, multiple years of conflict remove entire age cohorts from schooling and developing peacetime work skills, reducing employment options and future economic opportunity. In turn, a disenfranchised population and its subsequent generations are more likely to experience the negative synergism of poverty, hunger, disease, and the conditions that fueled past conflicts.

ARMED CONFLICT IN PERU: THE CASE OF SENDERO LUMINOSO

Beginning in 1980 and extending well into the 1990s, Sendero Luminoso (Shining Path) waged war against the Peruvian state and against all factions—even the political Left—who threatened its potential control of the countryside and its collaboration with peasant groups. The movement began in the impoverished Department of Ayacucho, under the leadership of philosophy professor Abimael Guzman and his followers. From this epicenter, it sought to expand its base in areas where poverty and vulnerability provided an opportunity for gaining recruits and resources to finance the revolution: cities, *selva* (remote areas, particularly coca-growing areas), and peasant communities of the southern sierra (McClintock 1989; Poole and Rénique 1992; Stern 1998).

The causes of conflict were many, but there is a general consensus that the seeds of revolt were sown in deep poverty, neglect, political marginalization, and little hope that future generations would fare any different (McClintock

1984; Poole and Rénique 1992). The history of this region and people is rooted in conquest and colonization, in postcolonial domination and exploitation of rural highland populations by a landed oligarchy, and by a series of more-recent shocks and crises. Peru, in many ways, is in perpetual crisis. Although per capita income in 1960 was the fifth highest in Latin America, so were levels of relative inequality. The percentage of households in poverty and chronic poverty was greater, and infant mortality was fourth highest in the region, 23% higher than the median for Latin America (Sheahan 1999). Between 1970 and 1990, the Peruvian population grew by 60%, but economic growth did not. In fact, in 1980 the rate of growth in gross domestic product was negative (United Nations 1996). Salaries dropped, unemployment and underemployment rose, inflation reached triple digits, real incomes were less than 50% of 1973 incomes (Reid 1985), and the poor became poorer.

In poor, rural, and largely indigenous communities, such as those in Ayacucho and Puno, levels of poverty, hunger, and illness were profound. Infant mortality rates in the southern sierra of the early 1980s were between 110 and 129 per 1,000, more than twice those of the city of Lima (Sheahan 1999). Estimates suggest that up to 60% of the population was in extreme poverty, and levels of undernutrition were equally high. Medical care was extremely limited, and while education programs had reduced illiteracy rates dramatically in Peru, illiteracy rates in the highlands remained five to nine times greater than those in Lima (Sheahan 1999). Moreover, industrial substitution policies of the central government led to stagnation of agriculture in rural highlands, and agrarian reform that was proposed to address land inequities was largely ineffective, further deepening the sense of hopelessness and anger at inaction and ill-conceived activities.

Once the war began in Ayacucho, a slow initial government response escalated to repressive military actions, in which atrocities on both sides led to death and displacement in many *campesino* (peasant) communities (McClintock 1989). The recent "Truth and Reconciliation Committee" estimated that 70 thousand people died during the 15 years of conflict, a casualty count approximately twice that of earlier government estimates. Likewise, despite earlier estimates that placed the number of displaced people at six hundred thousand, the true number is likely closer to one million (United Nations 1996).

A government report estimates that the material losses during conflict years amounted to US$21 billion, a figure equivalent to the country's entire foreign debt (United Nations 1996). Although hyperinflation of 7,000% declined to 15% in the mid-1990s, per capita gross domestic product was still at 1961 levels. In the 1990s, 16% of the Peruvian population earned less that one U.S. dollar per day (FAO 2002). The degree of food insecurity at the national level during conflict years is reflected in the percentage of undernourished, estimated by the FAO (2002) at 28% in 1980, 40% in 1991, and 11% in 1996. Health statistics in the highland indigenous zones did not worsen according to most national surveys, but neither did they improve dramatically. By contrast, more-recent surveys of health, nutrition, and infant mortality rates indicate improvement between the early and late 1990s. Levels of chronic undernutrition for children living in rural zones and the rural highlands (as indicated by their stunting or low height for age) were between 55% and 65% from the 1980s through the early 1990s but by 1996 had dropped to 40% (FAO 2000).

Impacts of conflict were felt most in rural indigenous communities in both highlands and lowlands. The majority of deaths and an estimated 70% of those displaced are from rural indigenous communities, especially the southern highlands from Ayacucho to Puno. The demographic structure there is now dominated by women and children, many of whom are orphans (up to 16% in some communities; United Nations 1996). Many of the displaced and returnees lost title to their land and often came home to face accusations of supporting Sendero Luminoso or conscription into the military or the peasant self-defense groups (*rondas campesinos*) that grew and flourished during the conflicts. Moreover, poverty and hunger still grip the countryside, and in some cases up to 80% of the displaced indigenous and native communities suffer from malnutrition. Insufficient health care in some areas (e.g., Ayacucho, Junin, Puno) has led to a persistence or an increase of cholera, tuberculosis, and other parasitic and infectious diseases. Respiratory disease, diarrhea, dysentery, and malaria plague large parts of the *selva*.

Social and psychological trauma (e.g., depression, nightmares, fear, distrust) may be the most widespread but unmeasured effect in rural areas. In addition, reports from the displaced and returnee communities cite that poverty and landlessness are increasingly associated with alcoholism, interpersonal and domestic violence, and "extremely aggressive" behavior of minors forcibly recruited by Sendero Luminoso or the *rondas campesinos* (United Nations 1996).

The View from Nuñoa

The causes and consequences of conflict in Peru are historically specific but not unlike those in many other places referenced in the previous section. The majority of reports on causes and consequences of armed conflict in Peru and throughout the world provides little insight into how these conditions are experienced at the household level, why some flee and some remain, and why some join the revolution while others do their best to ignore it or actively resist. In the following section, I attempt to provide added insight into the lives of the people experiencing the structural violence, poverty, hunger, and illness, those who are making "least worst" decisions about how to cope with these conditions and for whom little hope for change might well give way to actions that seek to overtly challenge systems of oppression (e.g., rustling, land invasions, or revolution). I conducted research into nutrition, health, food production, and household economy among rural producers in several communities differentially affected by agrarian reform and rural capitalism in the district of Nuñoa (Melgar, Puno) in the southern Peruvian Andes in 1983–1984, on the eve of heightened activity of Sendero Luminoso (Leatherman 1994, 1998). In many ways, the situation was desperate. The analyses pointed to the precarious position in which many households were located socially, economically, and in terms of household health and the ability to maintain the home and family as a unit of production and reproduction. It was clear at the time that the movement was growing and that it was only a matter of time before the violence made its way from Ayacucho and further north, into this part of the southern Andes.

The context for crisis in Puno, as elsewhere, was a combination of historical poverty, severe drought, and especially a feeling of further disillusionment and disenfranchisement over what was seen as a failed agrarian reform. The region of northern Puno was the site of many of the largest wool-producing haciendas. While most of the former haciendas were expropriated, most of the land was converted into large wool-producing cooperatives that employed only some of the workers from the original haciendas. Before the reform, 2% of the landholders in the district of Nuñoa controlled 61% of the land base; after the reform, three cooperatives controlled 60% of the land and employed only 25% of the rural population (Leatherman 1998). Very little land was turned over to peasant communities or private smallholders. Many rural households formerly linked to large landholders through sharecropping arrangements became landless, and there was especially a lack of access to land for small families to pasture sheep, llamas, or alpacas, denying such persons an important source of wealth, meat, and dung for fuel or fertilizer. As a result, towns grew rather rapidly with an influx of landless peasants looking for opportunities on the wage market or in petty commodity production. Yet local income-generating opportunities were limited, and wages paid were often below a living wage (e.g., one U.S. dollar per day). At the same time, wage and commodity markets expanded, as did the need for cash. Thus, most families in the region relied on a varied and complex set of food- and income-generating activities for their livelihood. These changing economic relations led to conflicts over the allocation of household labor to different economic tasks and the availability of extrahousehold labor to plant fields or help out when someone became ill.

In our research we studied three sites differentially affected by the reform and changing markets for goods and wage labor. One was an alpaca-herding cooperative of 25 families who received a wage for tending cooperative herds as well as ample usufruct rights to land for private herds and small-scale agriculture. A second was a small *ayllu* (traditional kin-based community) of 25 families that received no lands as part of the reform and found themselves being restricted on three sides by cooperatives that refused them access to pastures. Our third site was a semiurban town with a relatively diverse population of farmers, herders, shop owners, petty commodity producers, and landless unskilled laborers. Thus, while secure access to land for farming and herding benefited members of the alpaca-herding cooperative, access to land and resources for town and *ayllu* households remained the same or decreased following the reform. The *ayllu* particularly suffered because cooperatives discontinued informal arrangements through which some of these households accessed additional land for pasturing private herds and gathering resources such as dung for fuel and *ichu* grass for roofs. Indeed, when asked about the impacts of the agrarian reform on their community and households, 63% of cooperative families believed that conditions had improved, while only 5% of the *ayllu* saw improvements; 79% of this latter group thought that conditions had worsened. Almost 90% of the *ayllu* households reported owning less land and fewer animals than that of their parents. About half of the town households believed that conditions had improved or stayed the same, and half thought that they had worsened, especially landless or near-landless unskilled laborers.

Estimates of infant mortality, undernutrition, and illness illustrate the overall poor health in the region but were worse in the *ayllu* and in the poor landless (or near landless) in the semiurban town. Infant mortality in the district of Nuñoa was 128 per 1,000, one of the highest in the rural sierra, and rough estimates from the *ayllu* suggest rates closer to 200 per 1,000. Levels of undernutrition and illness were elevated in this region: 73% of children from the *ayllu* were considered chronically undernourished (stunted) and 34% severely stunted, compared to 54% and 16% to 19% in cooperative and town, respectively. Similarly, adults in the *ayllu* reported higher levels of illness (more cases and symptoms) and more work disruption due to illness. Households from the *ayllu* were estimated to lose about 75 days of adult labor each year due to illness, almost twice that of town households and three times that of the cooperative. Loss of adult labor is particularly important in Andean production systems that rely on strenuous physical work that often must be carried out in relatively narrow seasonal windows. Thus, failing to plant on time, ditch a field following a heavy rain, or harvest before hailstorms can lead to large losses in production.

In every community, higher levels of illness were associated with significant negative effects on production (50% fewer fields planted and harvested and 35% lowered productivity), reproducing the contexts of poverty and poor health and further heightening vulnerability (Leatherman 1998). The reproduction of poverty and poor health sent those most vulnerable into a spiral of household disintegration. Some sold their animals and land, or otherwise lost access to land, and were forced to move to the town or migrate out of the area. In other instances, impoverished households unable to care for their children would send them to live with middle-class townspeople as *criadas*, working as servants in exchange for food, clothing, and education. Thus, the situation for some was desperate, with little hope for improvement. The agrarian reform and the subsequent promises of further reallocation of land benefited a few and frustrated many.

By 1986, the threat of Sendero Luminoso was acutely felt in this region. First, a raid (*saqueo*) on cooperatives and larger landowners took place over several days, during which many herds were stolen and the homes of the landowners were vandalized and looted. These were not carried out by Sendero but by others who used its presence as a weapon and perhaps as inspiration. There is some local suspicion that many new entrepreneurs in the town acquired their capital for new businesses by profits they made in the *saqueo* and other, subsequent raids. When Sendero did make its presence felt, it was first through a public assassination of a supposed director (later realized as a case of mistaken identity) of the largest cooperative in the area. In three or four subsequent events, the town hall was demolished, police and other officials were killed, and the town was declared a "liberated zone." Retaliation by government forces put the area under siege by a second force, leaving most of the population caught in the middle.

It is not surprising that the targets of attack during the *saqueos* and the public assassinations were the cooperatives and wealthy landowners. Anger over profits of cooperatives and lack of benefit to local communities were among common topics of discourse in the early 1980s. People resented both the cooperatives' control of land and resources and the scarcities of local resources under cooperative control. For example, this region was a large meat producer, but relatively little was available locally. Some cooperatives made local peasants pay to cut *ichu* grass, which was commonly used for roofs and had been easily accessible in years past. Local cooperatives apparently were only spared further direct attacks by paying a protection fee. The government responded to the growth of Sendero and repeated demands of other peasant organizations by reallocating some lands from the cooperatives to recognized *communidades*. By 2003, many of the cooperatives had been dismantled or downsized, with more land further allocated to communities.

It is also not surprising that several of our key informants from the poorest and most marginalized communities were suspected as those highly involved in the *saqueo* and as supporters of *Sendero*. This possibility makes sense given the levels of vulnerability in their lives and little hope for change. Indeed, many members of the *ayllu* we studied moved to a new community and expanded their holdings (and have since done better). The current whereabouts of the former residents are not known. Perhaps, as elsewhere in Peru, they became the internally displaced who at least for the time being lost control of their land and former community.

For many townspeople, the period of Sendero's dominance (1989–1992) established a kind of fear and uncertainty (another sense of vulnerability) that is still felt and manifested in a reportedly heightened aura of wariness and distrust of other townspeople. Informants spoke of the "one thousand eyes and ears" of Sendero, which disciplined their speech. Between 1989 and 1992, inhabitants routinely locked their doors and turned off the lights in their homes as soon as night fell, hoping no one knocked and trying to avoid disturbances. Many fled the area and never returned, while others returned to lost holdings and attempted to rebuild. Some still reported stress and uneasiness, children who jumped at loud noises, and general worry about rumors of a Sendero resurgence.

The larger costs (or in some cases, benefits) of the revolution have yet to be established for this region. To date, no data have been reported on changing levels of illness during the height of the revolution. It is reasonable to expect that health problems became worse or at least showed no improvement given the already poor health and nutritional conditions that existed before 1986. It is also reasonable to expect that levels of production were affected given that Sendero was known to rob communities of their food, that *saqueos* and cattle rustling were more common, and that the flow of people and goods was more difficult. Since the early 1990s much land has shifted from cooperatives to communities; the town of Nuñoa has almost doubled in size; and more small shops and government enterprises are evident. There appears to be more wealth in the town, and there should be a more equitable allocation of land in the countryside. Yet the local discourse in the town during a recent visit in 2003 was that there was more poverty, hunger, and alcoholism in the town and rural communities. A recent survey of regional child growth (a common measure of community health) suggests no major changes in child growth between the 1980s and late 1990s (Pawson et al. 2001). During this same period, measures of nutrition on a national scale showed marked improvement (FAO 2002). Farming–herding production, illness-coping strategies, and resilience to the potentially devastating impact of illness on household livelihood are linked to the strength of social networks and interpersonal social relations (Leatherman 1998). Given a heightened wariness, distrust, and lack of cooperation in the community, we might expect that ability to cope with illness would decrease and that the impacts of poor health on household livelihood would be more severe.

It is impossible to say with certainty who sympathized with Sendero and who did not, who suffered most and who benefited most, or the extent of the social and health costs of being caught in the middle of a brutal revolutionary movement and equally brutal government response. These questions are for future research. The point is that the conditions that give rise to such events are tied to structures of poverty and inequality, and an analysis of a local and historically specific vulnerability can have something to say about these conditions.

CONCLUSION

Despite great strides in health and accumulation of wealth on a global scale, worldwide levels of poverty and inequality have not diminished. The poor carry the burden of a disproportionate share of malnutrition and disease and, in the last three decades, a disproportionate share of violence in armed conflicts. The contexts of conflicts are rooted in histories of chronic marginality, injustice, and exploitation and in current contexts of food insecurity, environmental struggles, and poor health. The consequences of armed conflicts often ensure the persistent vulnerability of populations due to the synergistic effects of death and displacement, degraded environments and food insecurity, and poverty and poor health. The forces of global capitalism that sever the linkages between producers and their products, combined with systems of unequal distribution, exacerbate these realities. Globalization also means that the tools of war are readily available, and the frequency with which they are brought to bear on marginalized groups increasingly makes violence a central context in which research on global health and environments must operate.

In this chapter I suggest that a political–ecological perspective can provide a suitable framework for research on armed conflict in that it explicitly frames analyses of environment, food insecurity, hunger, and violence within the interactions of structural inequalities and human agency. Unequal and inadequate access to environmental resources, whether through drought, degradation, or denial of rights, is a frequent corollary to food insecurity, and food insecurity is a precursor to hunger, malnutrition, and disease. The synergistic effects of malnutrition and disease are both an everyday reality of much of the world's poor and a major impediment to escaping poverty. Environmental scarcity and food insecurity are often evidence of impending conflict. But scarcity and insecurity are social and political constructions and thus must be viewed within broader contexts and loci of power—that is, who is in power with privileged access, and who is marginalized and denied access? Armed conflict and violence are always underlain by structural violence.

The question then becomes, in what contexts does structural violence become the collective violence of armed conflicts? This requires not only a sense of the historical and recent structural inequalities but attention to the perceptions, experiences, decisions, and actions (i.e., the agency) of the powerful and the weak. In the extreme threats on livelihood that normally form a part of the context of armed conflicts, there is always a range of response—some take up arms and some resist, some stay and some flee, some despair and some hope. The nature of these decisions and actions is key to understanding the course of conflicts and the impacts of conflicts on people's lives. I attempt to demonstrate within the discussion of the Sendero Luminoso revolt in southern Peru that close inspection of local contexts of vulnerability—failed agrarian reform, extremes of poverty and illness, and the potential for household disintegration—might get us closer to understanding the nature and variability of response. Thus, understanding the history of inequalities, the components of structural violence, the assaults on livelihood, and the individual and collective responses to these threats is essential for framing questions and analyses about the causes and consequences of armed conflicts.

REFERENCES

Blaike, P., and H. Brookfield
1987   Land Degradation and Society. London: Methuen.

Blim, M. L.
1992   Introduction: The Emerging Global Factory and Anthropology. *In* Anthropology and the Global Factory: Studies of the New Industrialization in the Late Twentieth Century. F. A. Rothstein and M. L. Blim, eds. Pp. 1–30. New York: Bergin and Garvey.

Bracken, Patrick, Joan Giller, and Derek Summerfield
1995   Psychological Responses to War and Atrocity: The Limitations of Current Concepts. Social Science and Medicine 40(8):1073–1082.

Bryant, R. L., and S. Bailey
1997   Third World Political Ecology. New York: Routledge.

Cervantes, R. C., V. N. Salgado, and A. M. Padilla
1989   Posttraumatic Stress in Immigrants from Central America and Mexico. Hospital and Community Psychiatry 40(6):615–619.

Chauvin, L.
2003   Peru: Truth, but No Reconciliation. World Press Review 55(11). Electronic document, from www.worldpress.org/Americas/1595.cfm, accessed on January 24, 2004.

Cohen, M. J., and P. Pinstrup-Andersen
1999   Food Security and Conflict. Social Research 66(1):375–416.

Donohue, Martin
2003   Causes and Health Consequences of Environmental Degradation and
        Social Injustice. Social Science and Medicine 56:573–587.

Farmer, Paul
1999   Infections and Inequalities: The Modern Plagues. Berkeley: University of
        California Press.
2003   Pathologies of Power: Health, Human Rights, and the New War on the
        Poor. Berkeley: University of California Press.

Feachem, R. G.
2000   Editorial: Poverty and Inequity: A Proper Focus for the New Century.
        Bulletin of the World Health Organization 78(1):1–2.

Food and Agriculture Organization
2000   Perfiles Nutricionales Por Paises: Peru. Rome: Food and Agriculture
        Organization.
2002   The State of Food Insecurity in the World, 2002. Rome: Food and
        Agriculture Organization.

Garrett, Laurie
1994   The Coming Plague: Newly Emerging Diseases in a World out of Balance.
        New York: Farrar, Straus & Giroux.
2000   Betrayal of Trust: The Collapse of Global Public Health. New York:
        Hyperion.

Gersham, John, and Alec Irwin
2000   Getting a Grip on the Global Economy. In Dying for Growth: Global
        Inequality and the Health of the Poor. Jim Yong Kim, Joyce Millen, Alec
        Irwin, and John Gersham, eds. Pp. 11–32. Monroe, ME: Common Courage
        Press.

Gleditsch, Nils Peter
1998   Armed Conflict and the Environment: A Critique of the Literature. Journal
        of Peace Research 35(3):381–400.

Heggenhougen, H. K.
1995   The Epidemiology of Functional Apartheid and Human Rights Abuses.
        Social Science and Medicine 40(3):281–284.

Hollifield, M., T. D. Warner, N. Lian, B. Krakow, J. H. Jenkins, J. Kesler, J. Stevenson, and J. Westermeyer
2002   Review: Measuring Trauma and Health Status in Refugees. Journal of the American Medical Association 288(5):611–621.

Homer-Dixon, T. F.
1994   Environmental Scarcities and Violent Conflict: Evidence from Cases. International Security 19(1):5–40.

Kalipeni, E., and J. Oppong
1998   The Refugee Crisis in Africa and Implications for Health and Disease: A Political Ecology Approach. Social Science and Medicine 46(12):1637–1653.

Kim, Jim Yong, Joyce Millen, Alec Irwin, and John Gersham
2000   Dying for Growth: Global Inequality and the Health of the Poor. Monroe, ME: Common Courage Press.

Latham, M. C., and M. Beaudry
2001   Globalization and Inequity as Determinants of Malnutrition: A Clear Need for Activism. Ecology of Food and Nutrition 40(6):597–617.

Leatherman, T. L.
1994   Health Implications of Changing Agrarian Economies in the Southern Andes. Human Organization 53(4):371–379.
1996   A Biocultural Perspective on Health and Household Economy in Southern Peru. Medical Anthropology Quarterly 10(4):476–495.
1998   Illness, Social Relations, and Household Production and Reproduction in the Andes of Southern Peru. In Building a New Biocultural Synthesis: Political-Economic Perspectives on Human Biology. A. Goodman and T. Leatherman, eds. Pp. 245–268. Ann Arbor: University of Michigan Press.

Leatherman, T. L., and R. B. Thomas
2001   Political Ecology and Constructions of Environment in Biological Anthropology. In New Directions in Anthropology and Environment: Intersections. C. Crumley, ed. Pp. 113–131. Walnut Creek, CA: AltaMira.

Le Billon, P.
2001   The Political Ecology of War: Natural Resources and Armed Conflicts. Political Geography 20:561–584.

Lundgren, R. I., and R. Lang
1989   "There is No Sea, Only Fish": Effects of United States Policy on the Health
        of the Displaced in El Salvador. Social Science and Medicine
        28(7):697–706.

Marmot, M., C. D. Ryff, L. L. Bumpass, M. Shiple, and N. F. Marks
1997   Social Inequities in Health: Next Questions and Converging Evidence.
        Social Science and Medicine 44(6):901–910.

McClintock, C.
1984   Why Peasants Rebel: The Case of Peru's Sendero Luminoso. World Politics
        37:48–84.
1989   Peru's Sendero Luminoso Rebellion: Origins and Trajectory. *In* Power
        and Popular Protest: Latin American Social Movements. S. Eckstein, ed.
        Pp. 61–101. Berkeley: University of California Press.

Messer, E., M. J. Cohen, and J. D'Costa
1998   Food from Peace: Breaking the Links between Conflict and Hunger. Food,
        Agriculture, and the Environment. Discussion Paper, 24. Washington, DC:
        International Food Policy Research Institute.

Millen, Joyce, Alec Irwin, and Jim Yong Kim
2000   Introduction: What Is Growing? Who Is Dying? *In* Dying for Growth:
        Global Inequality and the Health of the Poor. Jim Yong Kim, Joyce Millen,
        Alec Irwin, and John Gersham, eds. Pp. 3–10. Monroe, ME: Common
        Courage Press.

Pawson, Ivan G., Luis Huicho, Manuel Muro, and Alberto Pacheco
2001   Growth of Children in Two Economically Diverse Peruvian High-Altitude
        Communities. American Journal of Human Biology 13:323–340.

Peet, R., and M. Watts
1996   Liberation Ecologies: Environment, Development, Social Movements.
        New York: Routledge.

Poole, D., and G. Rénique
1992   Peru: Time of Fear. London: Latin American Bureau Limited.

Reid, M.
1985   Peru: Paths to Poverty. London: Latin American Bureau Limited.

Robbins, R. H.
2002   Global Problems and the Culture of Capitalism. Boston: Allyn and Bacon.

Sen, Aymarta
1993   Economics of Life and Death. Scientific American 268(5):40–47.
1999   Critical Reflection: Health in Development. Bulletin of the World Health
         Organization 77(8):619–623.

Sheahan, J.
1999   Searching for a Better Society: The Peruvian Economy from 1950.
         University Park: University of Pennsylvania Press.

Stern, Steve, ed.
1998   Shining and Other Paths: War and Society in Peru, 1980–1995. Durham,
         NC: Duke University Press.

Stockholm International Peace Research Institute
2002   SIPRI Yearbook 2002. Stockholm: Stockholm International Peace
         Research Institute.

Teodosijevic, Slobodanka
2003   Armed Conflicts and Food Security. ESA Working Paper, 03-11,
         Agricultural and Development Economics Division. Rome: Food and
         Agriculture Organization.

Ugalde, A., E. Selva-Sutter, C. Castillo, C. Paz, and S. Canas
2000   The Health Costs of War: Can They Be Measured? Lessons from
         El Salvador. British Medical Journal 321:169–172.

United Nations
1996   Profiles in Displacement: Peru. Report to Commission on Human Rights.
         Geneva, Switzerland: UN Commission on Human Rights.

United Nations Development Programme
1999   Human Development Report. New York: United Nations Development
         Programme.

Wallerstein, I.
1974   The Modern World System, Capitalist Agriculture and the Origins of the
         European World Economy in the Sixteenth Century. New York: Academic
         Press.

Watts, Michael
1983   Silent Violence: Food, Famine and Peasantry in Northern Nigeria. Berkeley:
         University of California Press.

World Health Organization
2002   Collective Violence. *In* World Report on Violence and Health. Pp. 214–239.
          Geneva: World Health Organization

Zwi, A., and A. Ugalde
1989   Towards an Epidemiological of Political Violence in the Third World. Social
          Science and Medicine 28(7):633–642.

II

CULTURAL ADAPTATIONS

4

Globalization, Migration, and Indigenous Commodification of Medicinal Plants in Chiapas, Mexico

David G. Casagrande

Traditional ethnomedical knowledge can be thought of as information that is necessary for survival and widely distributed throughout indigenous communities (Berlin and Berlin 1994; Garro 1986). It derives partially from direct experience with epidemiological and ecological realities, but it is also subject to cultural interpretations (Browner et al. 1988; Etkin 1988). Most important, traditional ethnomedical knowledge is put into practice through social cooperation (Barsh 1997). When I first began to study how medicinal plants are used in Tzeltal Maya communities of highland Chiapas, Mexico, my observations fit this description well. An elder man summed it up for me: "God made plants to cure every illness; it's our work to find them and give them to our families." I recorded many similar comments, and most people insisted that the medicinal plants that they use are crucial for saving lives, especially those of infants (Figure 4.1).

But ethnomedical knowledge is usually not distributed homogeneously throughout populations (see, e.g., Barrett 1995; Casagrande 2002; Garro 1986). Throughout the highlands of Chiapas, the Tzeltal Maya share a complex pharmacopoeia involving hundreds of plant species (Berlin and Berlin 1996:80). Within individual communities, there is a handful of plant species that are known for their medicinal use by nearly everyone; there is a set of about 40 to 60 such plants that are less widely known; and then there are hundreds of species known by only a few people (Berlin and Berlin 1996:82–87; Casagrande 2002:41–55; Stepp 1998). People tend to share this idiosyncratic

FIGURE 4.1
A Highland Tzeltal woman with her child. Many, if not most,
medicinal plant treatments pertain to infant survival.

information as needed. In this chapter I consider questions about what might happen to such an arrangement when confronted with the ideological and economic forces of globalization. How might the distribution of information change, and how might this affect a population's ability to cure illnesses?

During 2000 and 2001 I conducted a comparative study of Tzeltal in the temperate highlands and Tzeltal who had migrated from the highlands to the tropical lowlands, to see how neoliberal economic integration may have influenced patterns of knowledge distribution. While in the lowlands, I often heard comments about how medicinal plant knowledge had to be purchased. One man, desperate to find a cure for his child's infected fly bite, had traveled to another town and paid cash to a knowledgeable Tzeltal curer who taught him how to use a local tropical plant. The plant worked, but the man told me, "I'll never pay 300 pesos again. I decided to learn the plants myself." These types of comments were much more prevalent in the lowland migrant communities.

Two questions began to plague me: Why are the Tzeltal turning knowledge about medicinal plants into a commodity to sell to each other? And how might their doing so affect individual health? I was prepared to accept the injustice of pharmaceutical corporations' "pirating" knowledge and selling it back to indigenous people as pharmaceuticals, but this was different. A producer–consumer relationship appeared to be developing within the communities themselves. Surely some external forces are involved, but which? and how? And exactly how has migration opened the former reciprocal system to external influence?

In this chapter I describe a case of inchoate consumer capitalism in which control over personal health is shifting to those who accumulate capital within communities. This nascent hegemony is not imposed directly by external forces. The global processes that led to migration, combined with a new environmental context, have indirectly changed values and institutions of social organization so that some individuals are more likely to take advantage of capitalist opportunities and others are more likely to accept roles as consumers.

In the next section I briefly describe the research settings and methods used to obtain the data that support this argument. Then I describe the environmental constraints, traditional institutions, and forms of social organization that encourage the free circulation of medicinal plant knowledge in the highlands. I follow with a description of those global forces that have led to migration and an accelerated shift toward the commodification of medicinal knowledge in the lowland migrant communities.

Who wins in this case? It is probably too early to tell, but I conclude this chapter by suggesting possible effects of commodification on traditional knowledge and individual health.

THE STUDY SITES, METHODS, AND DEFINITIONS

The data I present here are from three communities located in two munici-palities. During the first six months, I carried out research in Nabil—a small community of about 150 households located in the municipality of Tenejapa at an elevation of 22 hundred meters (Figure 4.2). The vegetation is temper-ate, with mature stands of forest generally conforming to Dennis E. Breedlove's (1981) description of pine–oak–sweetgum communities. Ninety-nine percent of the people in Tenejapa are subsistence swidden (slash and burn) horticulturists (Instituto Nacional de Estadística, Geografía e Infor-mática [INEGI] 2001), primarily growing maize, beans, and squash. Frost, cold temperatures, and lack of rainfall preclude horticulture between Decem-ber and March in the higher, colder communities like Nabil. Residents of Nabil are part of a long history of patrilineally organized interaction with the nonhuman environment (Medina Hernández 1991). This history includes a shared pharmacopoeia that has developed over many generations.

During the subsequent six months I collected comparative data from the lowlands in two communities in Maravilla Tenejapa—a municipality founded within the last 30 years in an uninhabited tropical rain forest on the Guatemalan border by Tzeltal migrants from Tenejapa (Figure 4.2). I collected data within the municipal center and the small outlying community of Salto de Agua. These migrants speak Tzeltal and maintain most aspects of their subsis-tence economy, but they are also well integrated into the market economy. Al-though they still share many highland conceptualizations of illness, they have abandoned most traditional rituals and cosmology. The lowland communities have an elevation of about four hundred meters. The typical forest community is evergreen lower-montane tropical rainforest (Breedlove 1981), and the veg-etation is quite different from that of communities of origin in the highlands. The migrants have assembled a new pharmacopoeia within the last 30 years.

The primary language is Tzeltal in all three communities, and all surveys and interviews were conducted in Tzeltal. I used medical ethnobotanical sur-veys to document the distribution of medicinal plant knowledge in the three communities. I began by collecting and preparing interview specimens of all the plants that the residents might consider to be medicinal. In Nabil, this was based on research conducted in the highlands by my predecessors (Berlin et al. 1990; Stepp 1998). To this I added plants from 42 freelists that I elicited from adults, 10 four-hour trail surveys with five assistants, and interviews

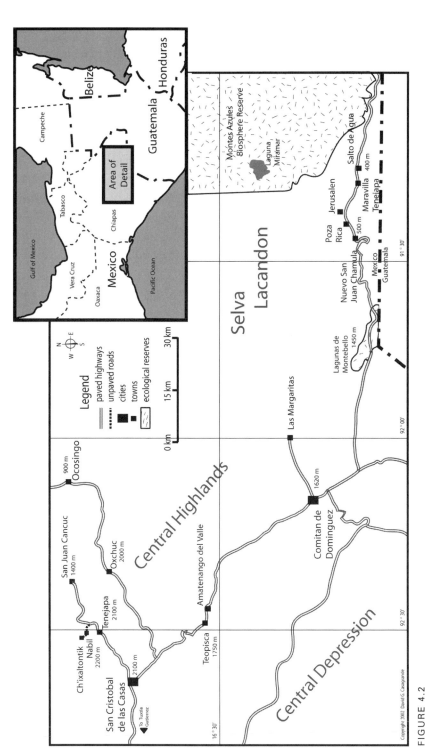

FIGURE 4.2
The study sites in Chiapas, Mexico.

with herbal shopkeepers. This yielded an interview collection of 257 potential medicinal species. I conducted structured interviews with 28 randomly selected adults (14 male, 14 female) who were over the age of 20. These interviews consisted of showing each specimen individually and asking interviewees for the plant's name, whether there was a medicinal use, and how they acquired their knowledge about the medicinal use. I conducted three door-to-door surveys (in May, June, and July 2001) to determine what illnesses existed in each of the 150 households in the past week and what actions were taken to cure the illnesses.

The comments I make here about changing social roles and social structure result from one-week time-allocation surveys of several households, a review of the ethnographic literature, analyses of tape-recorded discourses and narratives, and general observations I made while living in the communities and participating in activities such as religious ceremonies, planting and harvesting, and transporting harvested products to homes and markets.

I replicated the medical ethnobotanical interviews, household illness surveys, time-allocation surveys, and participant-observations in the lowland communities. Lowland ethnobotanical interviews were based on 116 potential medicinal species, and included 9 male and 9 female interviewees randomly selected in the Maravilla Tenejapa center, and 12 males and 12 females in Salto de Agua.

At this point it is important to clarify a few descriptive terms. Highland Tzeltal Maya who are recognized by their fellow Tzeltal as having some special talent for curing—usually due to supernatural powers—are linguistically labeled by the Tzeltal as *hposhiletik* (curers) (Brett 1994:48; Metzger and Williams 1963). Mayan–Catholic prayer and ritual are integrated into their curing practices. They are most likely to treat emotional illnesses and rely heavily on spiritual explanatory models to procure confidence. This is a prestige role. For their services, *hposhiletik* receive compensation in the form of meals, arranged travel, or other favors, but never cash payment. They often claim to have been called to their practice in a dream and command the authority and respect to prescribe specific behaviors. Most *hposhiletik* in the highlands today are elders, and I found little evidence of recruitment among younger generations. This suggests a diminishing role in the highlands. There are no *hposhiletik* in the predominantly Protestant migrant communities in the lowlands.

There are other specialists who receive monetary payment for their plant-based medicinal knowledge and who generally do not include supernatural, religious, or cosmological references in their practice. These people are considered by the Tzeltal to also be medicinal plant experts (Brett 1994:48–60) but are called *yierberos* (herbalists). The types of illnesses that they treat tend to have clear biomedical correlates, and they tend to employ biomedical explanatory models to justify their activities and win the confidence of patients (Casagrande 2002:6). They are likely to have invested financial resources into their practices and derive income by charging for services. *Yierberos* include the owners of herbal medicine shops and kiosks in markets but also an increasing number of Tzeltal who practice from their homes and have studied herbal medicine in some formal setting. There are some *yierberos* in the highlands. They are ubiquitous in the lowlands.

I refer to people who do not conform to either of the criteria mentioned as "novices." Some novices may command considerably more knowledge about medicinal plants than others and may be consulted for their advice about medicinal plants. But they are not called *hposhiletik* or *yierberos* by other Tzeltal, nor do they refer to themselves as such, because they do not conform to the descriptions stated here.

HIGHLAND RESISTANCE TO NEOLIBERAL ECONOMIC COMMODIFICATION

My goal in this chapter is to suggest explanations for why the Tzeltal are commodifying medicinal plant knowledge. In part, it results from neoliberal economic policies of the Mexican government to integrate indigenous populations into regional markets through the production of commodities such as maize or into international markets with products such as coffee (Collier 1994). Commodity production, of course, also allows external global commodities to penetrate Tzeltal ideology and patterns of exchange. My approach here is to explore possibilities for why the highland Tzeltal may be more resistant to this process. Later, I describe how processes of globalization in combination with those of migration have eroded these forms of resistance, allowing for greater economic integration in the lowlands and therefore commodification of indigenous knowledge. I begin by describing the distribution of knowledge of medicinal plants in the highlands.

During structured ethnobotanical surveys, I asked each interviewee how he or she had learned each of the treatments. Responses in Nabil indicated that most knowledge (75%) was transmitted among families and friends, most significantly from parents to children (Table 4.1). Knowledge was much less likely to be transmitted through financial transactions, such as through visiting *yierberos* and medicinal shops, purchasing books, or paying for a course. When combined, these sources accounted for 3% of learning events in Nabil. They accounted for 20% in the lowlands. Although the older migrants had some previous experience with tropical flora (mostly by working as wage laborers in lowland plantations), a large amount of new knowledge had to be acquired. (Note the extreme hesitancy to experiment.) Although 20% may seem a relatively small proportion of knowledge to purchase, this number represents some of the most serious cases of illness, ones in which all other treatments had failed. As illustrated in the anecdote about the man whose child was infected by a fly bite, capitalism has found its niche in desperation. Some knowledgeable novices, especially elders, continue to freely share information with friends and neighbors (Table 4.1), but no less than seven younger men in the lowlands told me that they were attempting to acquire information with the specific intent of establishing professional *yierbero* practices. This was based on their experiences of having to pay other *yierberos*. Thus, the trend toward commodification has

Table 4.1.   Sources of Individual Medicinal Plant Knowledge

	Nabil		Maravilla Tenejapa and Salto de Agua	
Information source	Reports (#)	%	Reports (#)	%
Parents	1,193	52.1	306	19.7
Other family members	530	23.2	204	13.1
Friends, neighbors, acquaintances	304	13.3	459	29.5
Traditional Tzeltal healers	31	1.4	5	0.3
Strangers	26	1.1	37	2.4
Markets, stores, traveling salespeople	26	1.1	78	5.0
Professional herbalists (yierberos)	24	1.0	88	5.7
Lowland plantations	22	1.0	109	7.0
Books	17	0.7	82	5.2
Biomedical professionals	16	0.7	54	3.5
Courses	11	0.5	61	3.9
Self-experimentation	4	0.2	5	0.3
Guatemalans	n/a	n/a	17	1.1
Don't remember	86	3.7	51	3.3
Total	2,290	100.0	1,556	100.0

Source: Structured ethnobotanical surveys.

begun. Why would some Tzeltal be willing to exploit knowledge so crucial to survival, especially for infants, and others acquiesce to this arrangement?

One major difference between the highlands and the lowlands is that the shared pharmacopoeia in the highlands predates the intrusion of consumer capitalism. Tenejapa was relatively inaccessible until the late 1970s, when it became connected to the nearby city of San Cristóbal and the Pan-American Highway. Electricity did not reach Nabil until 1994. Tourism and its affiliated home-based textile industry did not experience significant growth or economically influence remote areas such as Tenejapa until the 1970s. At the same time, large-scale government infrastructure projects that resulted from the oil boom of the late 1970s provided wage-labor opportunities in construction. This is not to say that Tenejapa was completely isolated from regional influences before 1970. Indeed, as George A. Collier (1975) argues, the patrilineal land-tenure system in the highlands is partly the result of regional forces, including capitalist intrusions. But the arrival of consumer capitalism is recent. Although Nike, Sony, and Tommy Hilfiger knockoffs can be seen in Tenejapa today, such consumption patterns were unheard of as recently as 25 years ago. Regarding medicinal plants, the people I interviewed who were born in the 1930s and 1940s identified many plants they learned as young children. Thus, the shared pharmacopoeia predates the arrival of global consumer capitalism.

On the other hand, the lowland pharmacopoeia has developed within the last 30 years concomitant with the spread of global consumer capitalist ideals. Indeed, the migration and its infrastructure are direct results of global economic and political intrusions beginning in the 1970s (Collier 1994). It seems only logical that the lowland system would be more "open" to commodification.

But this explanation is superficial at best. Now that global capitalism has arrived in the highlands, why wouldn't highland Tzeltal who are more knowledgeable about medicinal plants try to "cash in"? *Yierberos* are becoming more common in the highlands, but they mostly introduce new treatments learned elsewhere. Idiosyncratic traditional information held by *hposhiletik* and knowledgeable novices remains off-limits to the cash economy. Here, I suggest some reasons.

Environmental factors in high-elevation communities of Tenejapa provide poor opportunities for global economic integration. The cold weather in Nabil allows for only one maize crop to be grown per year and precludes cultivation of coffee and marketable tropical fruits, such as bananas, avocados, lemons, oranges, or mangoes. Tzeltal who live at lower elevations in Tenejapa can grow coffee, many tropical products, and two maize crops per year. Residents of Nabil have sought income from wage labor in nearby towns and cities, logging pine within Nabil and manufacturing and marketing crafts and textiles. But this has not been sufficient or reliable enough to replace the subsistence-based economy of swidden horticulture. Nearly everyone I spoke with said they would not forfeit their usufruct rights to patrilineally inherited land to invest more time in other economic pursuits. This presents a fundamental obstacle to neoliberal economic integration throughout Mexico. Meanwhile, environmental conditions and the unreliability of external income have precluded the ability to increase crop yields through investments in fertilizers, pesticides, or herbicides. As a result, residents of Nabil struggle to produce sufficient food for household consumption and are even less likely to produce for the market. Some households have acquired electronic appliances such as radios, but appliances such as gas stoves, televisions, or refrigerators are conspicuously absent when compared to households of lower-elevation communities. Only recently have home construction materials begun to change from wood walls and thatched roofing to cement block and aluminum (Figure 4.3).

Residents of Nabil are constantly at risk of malnutrition and disease, and they have little financial resources to deal with illnesses. Although children are vaccinated by the government, and some basic treatments are offered free of charge at the government health clinic in Tenejapa center and government hospitals in the nearby city of San Cristóbal, comprehensive biomedical care and pharmaceuticals are for the most part unaffordable to residents of Nabil. People often commented about their lack of ability to afford advanced treatments and pharmaceuticals or to pay for transportation costs to pursue these options. Household illness surveys clearly showed that locally known medicinal plants represent the most common strategy for treating illnesses (Table 4.2).

The household surveys in Nabil indicated high incidences of gastrointestinal and respiratory illnesses. This is consistent with reported diagnoses of visits to the government health clinic in Tenejapa (Table 4.3). Tenejapa has an infant mortality rate of 146 per 1,000 births (INEGI 2001). Most of the illness treatment events in Nabil households involve infants and children, and most of the plants in the shared pharmacopoeia target infant survival. My interviews in Nabil indicate that parents most often start with local plants. When

FIGURE 4.3
A Nabil kitchen building with pine-plank walls and thatched roof. In the wealthier high-land communities, thatched roofs have mostly been replaced by aluminum or as-bestos. Plank walls have been replaced by cement block.

symptoms do not respond, parents pursue other options, such as visits to clin-ics, pharmacies, or traditional healers. I suggest that open access to the shared pharmacopoeia in Nabil is ideologically and practically linked to the kin-based cooperative nature of subsistence in general. In the following, I argue that the epidemiological context remains the same for the lowlands but that institutions and the social organization of subsistence have changed. The low-land Tzeltal are more likely to pay for treatments first (Table 4.4).

Nabil is a closed patrilineal, patrilocal community. That is, land is passed from fathers to sons, and women live with their husbands' families. Thus, only through marriage can outsiders gain direct access to resources within the political bound-aries of Nabil. But marriages also create extended networks of labor reciprocity that are especially important during times of high labor demand, such as when planting maize. From a practical standpoint, most medicinal plant information

Table 4.2. Illness Frequencies by Class, and Actions Taken, as Reported by Residents of Nabil (Highlands)

Illness Category	Local Plants	Recourse				Pharmacy	No Action	Total
		Hposhiletik	Yierbero	Clinic	Hospital			
Gastrointestinal	88	0	0	11	0	3	25	127
Respiratory	63	0	1	13	0	20	18	115
Dermatological	6	0	0	7	0	19	0	32
Fever	13	0	0	0	0	5	0	18
Headache	13	0	0	0	0	1	0	14
Body aches	19	1	1	0	0	4	24	49
Oral thrush	0	0	0	0	0	0	0	0
Reproductive	0	0	0	0	1	0	0	1
Emotional/psychological	0	2	0	0	0	0	0	2
Dental	5	0	0	0	0	0	0	5
Total	207	3	2	31	1	52	67	363

Note: *hposhiletik* = traditional healer, *yierbero* = professional herbalist, clinic = biomedical.

Table 4.3.   Cases of Visits to the Instituto
Mexicano del Seguro Social Health Clinic in
Tenejapa Center During 1999 and 2000

Diagnosis	No. Cases
Respiratory infections	283
Whooping cough	0
Pneumonia	0
Tuberculosis	0
Total respiratory	283
Intestinal amoebas	36
Ascaris	6
Gastritis	15
Acute diarrheas	55
Total gastrointestinal	112
Urinary infections	28
Malnutrition	17
Conjunctivitis	9
Scabies	7
Ear inflammations	7
Diabetes	6
Vaginal infection	2
Neurological shock	2
Dog bites	1
Cervical dysplasia	1
Chronic alcoholism	1
Measles	0
Tetanus	0
Total miscellaneous	81
Total	476

Source: Instituto Mexicano del Seguro Social annual summaries of daily register of consultations.

is exchanged within families, especially from parents to children (Table 4.1). Because of marriages, knowledge is also exchanged across political boundaries and ecological zones. A crucial aspect of transmission is the issue of trust in the legitimacy of information. My analyses of Tzeltal medical discourses have shown that both highland and lowland Tzeltal are extremely concerned about the safety of ingesting plants and carefully weigh the legitimacy of information sources (Casagrande 2002:185). This is also illustrated by the extreme reluctance to self-experiment with new medicinal plants and a tendency to rely more on information from family and friends (Table 4.1). Survival and safety, including both subsistence-based food production and infant care, are practically and ideologically linked by kin-based trust and cooperation. These relationships are not compatible with migration.

Table 4.4. Illness Frequencies by Class, and Actions Taken, as Reported by Residents of Maravilla Tenejapa and Salto de Agua (Lowlands)

Illness category	Local Plants	Hposhiletik	Yierbero	Clinic	Hospital	Pharmacy	No Action	Total
				Recourse				
Gastrointestinal	88	0	0	11	0	3	25	127
Respiratory	31	0	7	19	0	27	10	94
Dermatological	9	0	4	4	0	16	0	33
Fever	8	0	3	3	0	5	1	20
Headache	9	0	1	0	0	3	0	13
Body aches	21	0	5	1	0	8	32	67
Oral thrush	5	0	3	3	0	2	0	13
Reproductive	17	0	8	0	0	0	7	32
Emotional/psychological	2	0	5	0	0	0	0	7
Dental	2	0	1	0	0	0	0	3
Total	165	0	52	59	0	70	80	426

Note: *hposhiletik* = traditional healer, *yierbero* = professional herbalist, clinic = biomedical.

Historical discourses that I recorded in the migrant communities describe two types of people who chose to migrate: risk-taking leaders and those who were so thoroughly marginalized in their communities that they had little to lose (see also Calvo Sánchez et al. 1989). Those who chose not to migrate preferred not to take the risk of leaving their cooperative networks.

GLOBAL PROCESSES AND MIGRATION

The Tzeltal migration from the highlands to Maravilla Tenejapa was clearly a result of global processes—in particular, neoliberal economic policies required for the Mexican government to restructure its foreign debt. Rising oil revenues in the mid-1970s enabled the federal government to undertake major capital investments in Chiapas, including highways and hydroelectric dams. These created opportunities for wage labor in construction for many highland Maya (Collier 1994), which in turn contributed to an increase in fertility. In the meantime, education and immunization programs for polio, measles, whooping cough, diphtheria, and neonatal tetanus were reducing infant mortality. Population density in Tenejapa rose to an estimated 56 per square kilometer in 1974 (Berlin et al. 1974)—a figure that could not be supported by the traditional maize-swidden system (Collier 1975). Migrants that I interviewed strongly believed that there was simply not enough land of sufficient quality to support their families.

The government planned to relieve population pressure by building a road through the uninhabited area of the Lacandon Rain Forest along the Guatemalan border (see Figure 4.2). The goal was to extract resources and encourage settlement of the area by migrants from the highlands (Calvo Sánchez et al. 1989; Collier 1994). The urgency of this plan increased with the escalation of the war in Guatemala—a war that was largely financed by the United States and the Soviet Union. The Mexican government hoped to use the road to secure the border with troops and farmers. But the road was also a symbol of neoliberal economic policies dating back to the 1950s. Opening the Lacandon Rain Forest to logging and oil exploration in attempts to reduce mounting foreign debt after the price of oil began to fall went hand-in-hand with attempts to integrate indigenous populations into international markets through the production of commodities such as coffee.

The Mexican government began to build the road during the late 1950s. In 1973, the government announced that there was still land available for

settlement, and 35 families from various communities in Tenejapa, including Nabil, formed a group who petitioned the government for a land grant.
This group founded Maravilla Tenejapa in 1974. In 1987, 34 families from
Tenejapa and Maravilla Tenejapa founded a satellite community approximately seven kilometers to the east called Salto de Agua (see Figure 4.2).
Maravilla Tenejapa was actually founded seven years before the road reached
its location. The gravel road arrived in 1981 and was paved in 1996.

Today, most residents of the migrant communities near the road are not
only able to grow sufficient food for household consumption but can also
generate a modest income from the small-scale farming of corn, beans, coffee,
cattle, bananas, pineapples, and other tropical fruits. The government's plan
appears to have been successful. For better or for worse, the lowland Tzeltal are
more integrated, economically and ideologically, with the global capitalist system than are their counterparts whom I lived with in the highlands.

ETHNOMEDICINE IN THE MIGRANT LOWLANDS

Carving a life out of the rain forest was painstakingly difficult. Many migrants
gave up and returned to their homes. Diseases were particularly devastating.
The migrants had to cope with gastrointestinal and respiratory illnesses the
same as in the highlands. Several informants told me that dysentery was ram-
pant. In addition, they had to deal with new tropical diseases, such as malaria
and dengue, and an increase in emotional illnesses. Worse yet, they had a poor
understanding of how to use the local flora as medicinals. Comparative floris-
tic surveys that I conducted in Nabil and Maravilla Tenejapa revealed an 81%
difference in botanical species. Although some migrants had experience with
tropical medicinal flora, most plants were new to them. Before the road
reached Maravilla Tenejapa, a trip to the nearest government clinic required a
two-day walk. Many of the migrant communities were in dire need of medical information. It was the *yierberos*, the professional herbalists, who first met
this demand.

Processes that led to the proliferation of *yierberos* in the lowlands can be
traced back to the highlands and are integrally linked with global economic
processes—in this case, European capital and left-wing ideology. Indigenous
healers in the highlands have formed a variety of political organizations in response to economic opportunity and perceived threats to traditional knowledge. Two of the most important are OMIECH and COMPITCH—in English,

the Organization of Indigenous Healers in the Highlands and the Council of Traditional Indigenous Doctors and Midwives of Chiapas, respectively. These organizations sponsor workshops, accept donations, and derive revenues from the sale of traditional remedies to other nonindigenous Maya Mexicans and predominantly European tourists. With the help of direct grants from European institutions, OMIECH has opened a museum, research, and retail center in the highland city of San Cristóbal.

European capital flowing into a system in which some people have the ability to accumulate and control capital without regulation, albeit with the best of intentions, is very likely to stimulate economic stratification. The indigenous healers are often portrayed as beneficent stewards of traditional knowledge (Nigh 2002). This is, for the most part, true. But their Tzeltal clients whom I spoke with are less than enthusiastic about paying cash for services that were, until recently, acquired on a basis of reciprocity. The bottom line is that a few Maya are accumulating or otherwise controlling capital via the transformation of "traditional" knowledge. European tourists or philanthropists and journalists on brief fact-finding visits are not privy to the details of such a system.

During the summer of 1998, I worked closely with a *yierbero* in Tenejapa who had paid to receive training by COMPITCH and OMIECH. He told me how the knowledge he acquired came from the original communities of his teachers, nonindigenous Mexican teachers, European herbalists who gave workshops, and books. I was impressed with his claims that some of the information came from scientific literature. In a few cases I was able to verify this. Most important, he told me that his teachers had visited Maya communities to learn new treatments—in essence, to conduct field research. There are three important points to this story. First, these organizations are acquiring knowledge in one place and marketing it in another. Second, capital investment is stimulating the development of more "traditional knowledge." Third, through the ability of some individuals to control capital, knowledge is transformed into a commodity. I should point out that many of the traditional healers in these organizations do not derive income for their services and many adhere to traditional cosmological ideologies. Nevertheless, they are fueling the growth of a side industry of *yierberos*. This process appears to be more pronounced in the lowland migrant communities.

Tenejapa, in the highlands, has a population of about 33 thousand (INEGI 2001). While living in the commercial and political center, I attempted to locate as many *yierberos* as possible. I found six in Tenejapa's center and two in the out-lying communities. Maravilla Tenejapa has a population of 10,500. Without try-ing, I met no less than 20 *yierberos* in this lowland community within a few weeks.

The proliferation of *yierberos* in the lowlands is not only a function of cap-ital investment but also demand. And demand in the lowlands is a function of both medical needs and an increased ability to pay for services, especially when compared to that in a highland community such as Nabil. Residents in Maravilla Tenejapa and Salto de Agua are able to grow two, and sometimes three, maize crops per year using traditional swidden techniques. In addition, they grow coffee and other tropical crops for household and market con-sumption. They are also probably growing illegal crops for the global nar-cotics industry that permeates the region. Access to the road—itself a result of global economics, war, and neoliberal economic policies—is crucial for these farmers to generate income from surplus production.

But economic processes are insufficient for explaining the widespread ac-ceptance of a cash-based, producer–consumer relationship that appears to meet greater resistance in the highlands. Consumer demand and the willing-ness to pay other Maya for information reflect shifts in identity, values, and ideals. Here, I provide a few examples.

One of the most striking differences between Tenejapa and lowland Mar-avilla Tenejapa is that social institutions such as civil–religious *cargos* and Mayan–Catholic *hposhiletik* do not exist in Maravilla Tenejapa. The *cargo* sys-tem is a hierarchical arrangement of ranked, alternating civil and religious of-fices that all men in a community such as Tenejapa are expected to ascend in their lifetimes. The *cargo* system is ubiquitous throughout highland Chiapas. Some anthropologists have proposed that *cargos* preclude individuals from accumulating too much capital, while simultaneously reifying ethnicity (Can-cian 1965; Medina Hernández 1991; Wolf 1957). These offices place a consid-erable financial burden on each man, but they also confer prestige (Cancian 1965). The ability to acquire prestige in highland communities like Tenejapa was once limited to the roles of *cargos* and *hposhiletik* (Medina Hernández 1991). Frank Cancian (1992) describes how the *cargo* system has broken down in Zinacantán—a nearby Tzotzil municipality. In the 1960s community ser-vice and *cargo* service were synonymous in Zinacantán, and *cargos* were the most important marker of status. *Cargos* redistributed capital accumulated by individuals and transformed that capital into individual and family prestige. Greater regional economic integration in the 1970s opened alternative paths to social position and influence. Despite organized resistance, Zinacantecos gradually came to accept the continuous accumulation of capital by individuals. Similar processes throughout the highlands have resulted in significant socioeconomic stratification within indigenous communities (Collier 1994). In Tenejapa, bragging or displays of wealth were previously socially discouraged, a situation that would impede commodification of shared knowledge. Communities such as Zinacantan and Tenejapa are now in transition from civil–religious to class-based prestige (Cancian 1992). In the migrant lowlands, this transition is complete.

One explanation for leaving the highlands was to escape the marginalizing effects of social institutions such as the *cargo* system (Collier 1994). This, combined with the dominance of Protestant evangelism in the lowlands, precluded the re-creation of *cargos* in the new communities. Likewise, there are no *hposhiletik*. The only options for accumulating prestige are class based and rely on differential capacities to accumulate capital. In an environment in which money has replaced reciprocity, and prestige is based on accumulation of capital, charging for knowledge that was once common property has become acceptable.

The transition from civil–religious to class-based prestige and mobilization of capital obscures differences between the Tzeltal and nonindigenous Mexicans, thus eroding ethnic identity. Furthermore, community membership in the migrant communities is not patrilineal. Most migrant communities were formed by people from different towns in the highlands. In some cases, they include Tzeltal, Tzotzil, and Spanish speakers. There are clear socioeconomic strata, including the ability for Tzeltal landowners to hire itinerant Guatemalan laborers. Since most families are removed from their kin, there is no opportunity for kin-based reciprocity. The migrants have become comfortable with paying for help.

It is a small step from replacing ethnic identity with class identity to developing a consumer identity. Consumer goods such as refrigerators and televisions are far more common in the lowlands than in Nabil. Most houses have direct-television satellite subscriptions (Figure 4.4). Not only is wealth more prominently displayed, but the migrants have a greater exposure to consumer ideals delivered by media such as television.

In short, not only do the migrants have more money to pay for herbal treatments than do residents of Nabil, but they are also more willing to do so. Household illness and activity surveys I conducted in the lowlands indicate that the migrants are more likely to pay *yierberos* for knowledge, more likely to visit clinics than in the highlands, and more likely to purchase pharmaceuticals (Table 4.4). In addition, they are more likely to begin treating infants with these options than to begin with home-based herbal remedies.

The accelerated commodification of medicinal plant knowledge in the lowlands has not necessarily reduced the overall amount of "traditional knowledge." Indeed, the diversion of capital may have stimulated an increase in knowledge. The quality of that knowledge, however, is another issue. The knowledge in the highlands has been subject to many generations of trial and error, and its transmission is "filtered" by concerns about legitimacy and safety. Much of the information in the lowlands is only beginning to be tested in new epidemiological, ecological, and cultural contexts. On several occasions I witnessed transmission errors, such as people mistaking nonlocal botanical species seen in books for local plants. I also worry about the effects

FIGURE 4.4
Most Tzeltal households in Maravilla Tenejapa have consumer amenities such as satellite television.

of information being produced by the Tzeltal primarily for the sake of accumulating more capital the same as I worry about the motivations of unregulated multinational pharmaceutical firms.

Although commodification may have encouraged development of knowledge, it is more concentrated among fewer individuals than it is in the highlands. The mean number of medicinal plants known by individuals in Nabil is 47.7 ($SD = 26.1$). Some people knew over 100 plants. Some knew fewer, but I never found anyone over 12 years old who did not know at least 20 plants. In Maravilla Tenejapa and Salto de Agua, the mean number of medicinal plants known was 28.9 ($SD = 27.8$). Fewer plants are known (on average) by individuals in the lowlands. More important is the high standard deviation, which indicates a high degree of divergence between those who knew many plants and those who knew few. To my amazement, I found several individuals who could not provide the medicinal uses of any plants. In contrast to the general population, lowland *yierberos* could name hundreds of uses.

The effects of this asymmetrical knowledge distribution on individual health are unclear and deserve further study. Because malaria and dengue have been eradicated through intensive government programs, the lowland Tzeltal appear to be experiencing health problems similar to those in Nabil (Tables 4.2 and 4.4). Nevertheless, infant mortality in Maravilla Tenejapa (101 per 1,000 live births; INEGI 2001) is lower than that in Tenejapa (146 per 1,000 live births). Many factors may be contributing to improved infant survival in the lowlands. Nutrition is better because agriculture is more productive, reliable, and diverse. The government health clinic in Maravilla Tenejapa is more regularly staffed and better supplied than in Tenejapa (at least since the Zapatista uprising). Government programs and indigenous cooperatives have greatly improved sanitation. Every house in Maravilla Tenejapa has a cement latrine, while hardly any house in Nabil has any type of latrine. In addition, most lowland migrants appear to have a greater capacity to pay for pharmaceutical and herbal treatments, although this is not true for all migrants.

It remains unclear whether individual health really has improved, and it is even less clear what the contributions of *yierberos* have been. Most of the improvements in the lowlands have come at a financial cost to individuals in an atmosphere of increasingly asymmetrical capacities to accumulate capital. Surprisingly, some interviewees told me it was less expensive to buy pharmaceuticals than to visit *yierberos*. Also, there are some migrants who have not been successful and cannot afford medicines of either type. Even the poorest people in Nabil still have access to a well-tested, kin-based system of shared ethnomedical knowledge.

CONCLUSION

Global processes led to the migration from Tenejapa to the lowlands, thus changing values, identity, and social organization in the process. Control over personal health in this context is shifting to those who accumulate capital within communities. What was once free information available to all is becoming concentrated in the hands of fewer individuals and transformed into a commodity. A commodity perceived as a necessity and controlled by a privileged few provides dangerous opportunities for the few to accumulate even more capital and power.

Organizations of traditional curers have already used this power to their advantage. Beginning in 1998, an international bioprospecting consortium attempted to bypass traditional curers, preferring instead to test widely distributed traditional knowledge for pharmaceutical potential (Berlin and Berlin 2004; Nigh 2002). Their goal was to distribute financial compensation for this knowledge throughout the communities. Sensing a threat to their interests, traditional curers allied themselves with international opponents of bioprospecting to denounce the project, which was eventually cancelled. The lesson for future bioprospecting in Chiapas was clear. Any compensation must flow through those indigenous individuals who control knowledge and capital.

All of this leaves me with more disturbing questions than I started with. Do the seeds of hegemony, so often attributed to global interests, already exist within indigenous communities? Who wins in a situation where traditional knowledge has been transformed into a commodity by indigenous people within their own communities?

REFERENCES

Barrett, Bruce
1995   Herbal Knowledge on Nicaragua Atlantic Coast: Consensus within
        Diversity. Journal of Community Health 20(5):403–421.

Barsh, Russel
1997   The Epistemology of Traditional Healing Systems. Human Organization
        56(1):28–37.

Berlin, Brent, and Elois Ann Berlin
1994   Anthropological Issues in Medical Ethnobotany. *In* Ethnobotany and the
        Search for New Drugs. D. J. Chadwick and J. Marsh, eds. Pp. 240–265. Ciba
        Foundation Symposium, 185. New York: Wiley.
2004   Community Autonomy and the Maya ICBG Project in Chiapas, Mexico:
        How a Bio-prospecting Project That Should Have Succeeded Failed.
        Human Organization 63(4):472–486.

Berlin, Brent, Elois Ann Berlin, Dennis E. Breedlove, Thomas Duncan, Victor M. Jara
    Astorga, Robert M. Laughlin, and Teresa Velasco Castañeda
1990   La Herbolaria Médica Tzeltal-Tzotzil en los Altos de Chiapas: Un Ensayo
        Preliminar sobre las Cincuenta Especies Botánicas de Uso Más Frecuente.
        Volume 1. Tuxtla Gutiérrrez, Chiapas, Mexico: Instituto Chiapaneco
        de Cultura.

Berlin, Brent, Dennis E. Breedlove, and Peter H. Raven
1974   Principles of Tzeltal Plant Classification. New York: Academic Press.

Berlin, Elois Ann, and Brent Berlin
1996   Medical Ethnobiology of the Highland Maya of Chiapas, Mexico:
        The Gastrointestinal Diseases. Princeton, NJ: Princeton University Press.

Breedlove, Dennis E.
1981   Part 1: Introduction to the Flora of Chiapas. San Francisco:
        California Academy of Sciences.

Brett, John A.
1994   Medicinal Plant Selection Criteria among the Tzeltal Maya of Highland
        Chiapas, Mexico. Ph.D. dissertation, University of California.

Browner, C. H., Bernard R. Ortiz de Montellano, and Arthur J. Rubel
1988   A Methodology for Cross-Cultural Ethnomedical Research. Current
        Anthropology 29(5):681–702.

Calvo Sánchez, Angelino, Anna María Garza Caligaris, María Fernanda Paz Salinas,
    and Luana María Ruiz Ortiz
1989   Voces de la Historia: Nuevo San Juan Chamula, Nuevo Huixtán, Nuevo
        Matsam. San Cristobal de Las Casas, Chiapas, Mexico: Centro de Estudios
        Indígenas, Universidad Autónoma de Chiapas.

Cancian, Frank
1965   Economics and Prestige in a Maya Community: The Religious Cargo
        System in Zinacantan. Stanford: Stanford University Press.

1992   The Decline of Community in Zinacantan: Economy, Public Life, and
          Social Stratification, 1960–1987. Stanford: Stanford University Press.

Casagrande, David G.
2002   Ecology, Cognition, and Cultural Transmission of Tzeltal Maya Medicinal
          Plant Knowledge. Ph.D. dissertation, University of Georgia.

Collier, George A.
1975   Fields of the Tzotzil: The Ecological Bases of Tradition in Highland
          Chiapas. Austin: University of Texas Press.
1994   Roots of the Rebellion in Chiapas. Cultural Survival Quarterly 18(1):14–18.

Etkin, Nina L.
1988   Cultural Constructions of Efficacy. In The Context of Medicines in
          Developing Countries. S. van der Geest and R. Whyte, eds. Pp. 299–326.
          New York: Kluwer.

Garro, Linda C.
1986   Intracultural Variation in Folk Medical Knowledge: A Comparison between
          Curers and Noncurers. American Anthropologist 88:351–370.

Instituto Nacional de Estadística, Geografía e Informática
2001   El XII Censo General de Población y Vivienda 2000. Aguascalientes:
          Instituto Nacional de Estadística, Geografía e Informática.

Medina Hernández, Andrés
1991   Tenejapa: Familia y Tradición en un Pueblo Tzeltal. Tuxtla Gutiérrez, Chiapas,
          Mexico: Gobierno del Estado de Chiapas, Instituto Chiapaneco de Cultura.

Metzger, Duane, and Gerald Williams
1963   Tenejapa Medicine I: The Curer. Southwestern Journal of Anthropology
          19:216–234.

Nigh, Ronald
2002   Maya Medicine in the Biological Gaze: Bioprospecting Research as Herbal
          Fetishism. Current Anthropology 43(3):451–476.

Stepp, John R.
1998   Ethnobiología en los Altos de Chiapas: Una Revisión de la Distribución
          de las Plantas Medicinales. Geografia Aplicada y Desarrollo, Quito, Ecuador,
          1998. Vol. 18, pp. 52–62. Centro Panamerica de Estudios e Investigaciones
          Geográficas.

Wolf, Eric R.
1957   Closed Corporate Peasant Communities in Mesoamerica and Central Java.
          Southwestern Journal of Anthropology 13(1):1–15.

5

Health Ecology in Nunavut: Inuit Elders' Concepts of Nutrition, Health, and Political Change

Ann McElroy

In this chapter, I discuss the health ecology of political transformations in Canadian Inuit communities. On April 1, 1999, 26 communities previously part of the Northwest Territories established a two-million-square-kilometer territory called Nunavut, meaning "our land" in the language of Inuit. Still politically part of Canada, the new territorial government is founded on Canadian parliamentary and democratic principles; uses consensus governance; and follows principles of "maximum cooperation, effective use of leadership resources, and common accountability" (Government of Nunavut 2004). In this age of globalization, when cultural distinctiveness is often reduced to exotic stereotypes and nostalgia for the past, Nunavut stands as a model of ethnic revitalization and cultural persistence.

Macropolitical forces underlying Nunavut's creation are linked to a history of ecosystemic change. Change was initially due to local impacts of 19th-century contacts with European whalers and traders, but in the 20th century environmental change in the eastern Arctic ensued from increasingly wider influences: toxic waste and air pollution, climate change, and shifting markets for marine mammal products. The framework used here for analysis of the impacts of these transformations is health ecology, the study of cultural, environmental, and biological variables as a synergistic system affecting health. In health ecology, community access to resources is a principal variable, whether facilitated or constrained by political forces acting on regional and local ecologies (McElroy 1996).

Inuit bands of hunters traditionally subsisted on maritime and inland resources in the eastern Arctic in ways that minimized environmental degradation. Population density was low, about 0.03 persons per square kilometer,
and community size reflected a seasonal fission–fusion pattern, ranging from
12 to 70 residents per camp. The diet was low in carbohydrates and high in
protein, primarily from seal, caribou, whale, and fish. Although high in fats,
the omega-3 fatty acids in seal, whale, fish, and bear meat contributed to good
cardiovascular health, as did the activity level of the foraging lifestyle. Dietary
health problems were primarily confined to botulism, excessive vitamin A,
trichinosis and other parasitic infections, and osteoporosis. Diabetes, heart
disease, and cancer were rare.

The earliest documented Inuit contact with Europeans occurred between
1576 and 1578, when Martin Frobisher's English fleet sailed up the bay that
was later named after him. He sought a northwest passage to the East. Interactions between sailors and Inuit were violent; hostages were taken; and several people were killed (Hall 1970). Health ecology began to change shortly
after a series of 19th-century contacts with Europeans and later after consoli-
dation of settlements by Canadian authorities (Boas 1888; Eber 1989; Pitseolak and Eber 1975; Stevenson 1997). Since the 1980s, the animal rights
movement has added another chapter in the history of adverse economic consequences of outsiders' values and policies on indigenous communities. Seeking to conserve marine mammal populations, activists have sought to limit
export of marine animal products and to promote quotas on threatened
species, an agenda that has also constrained Inuit access to traditional foods
(Wenzel 1991).

Politicization of natural resources has been intrinsically at odds with cultural ideology. Inuit traditionally regarded the animals they hunted as sen-
tient, intelligent beings that shared the Inuit's environment and deserved their
respect (Wenzel 1991:1348). A respectful attitude meant that the hunter
would use the meat of game animals as food to be shared by all. Animal flesh
and blood, particularly from seals, was regarded as being essential for health
and strength (Borré 1991:54). These attitudes have persisted throughout
decades of modernization and in spite of stereotyped distortions of Inuit values by animal rights organizations.

Canadian policies have favored centralization of indigenous groups, primarily for efficient and equitable administration (Dickerson 1992). This pol-

icy was essential for addressing poor health conditions among Inuit during the early 20th century that were due to epidemics of introduced contagious diseases and nutritional insufficiency of trade foods (Hankins 2000). There is some debate whether people were propelled by external pressures, such as Royal Canadian Mounted Police (RCMP) policy and school regulations, to settle in towns or whether they were moving voluntarily to seek more secure access to food, supplies, and medical care. One could argue that transformations from nomadic foraging to wage labor in towns were pragmatic decisions reflecting the adaptive flexibility of Inuit. The town environment, with its many lifelines to southern Canadian technology and support systems so critical in times of hunger and illness, may be considered an expedient alternative to living on the land.

Political economists, on the other hand, regard transformations in settlement pattern as a product of habitat destruction, coercive policies, and exploitation by intruders seeking profit and control. Centralized administration, a cash-based economy, and dominance of European institutions are viewed as predictable stages in the loss of cultural integrity among many indigenous people. These stages of change are intrinsic to pressures reaching even the smallest and most isolated arctic villages. Attempts by Inuit to resist change by returning to independent subsistence on the land have often been short-lived, especially when illness and hunting misadventures occur (see Honigmann and Honigmann 1965:11–18 for a case in Iqaluit in 1962).

In this chapter, I examine the changes that Inuit have undergone, and I do so within a health ecology perspective, discussing issues associated with the Inuit's greater integration into the global community. Mobility, subsistence, diet, and identity are focal points of my discussion. My analysis draws on my ethnographic data from 1967 to 2002; interviews with Inuit elders in 1999 and 2002; and both published and unpublished materials on Baffin Island history, on the development of Nunavut, and on health changes in arctic communities. The chronology of political and ecological transformations in this chapter is accompanied by excerpts from Inuit life stories to illustrate the human impacts of colonialism and globalization.

STUDY METHODOLOGY AND RESEARCH SITES

Ethnohistory provides clues for reconstructing the interpersonal dynamics of cultural change. Inuit elders are a valuable repository for information about events in the decades of 1920 to 1980. My research on elders' perceptions of culture change is part of a longitudinal study of two Baffin Island communities, Iqaluit and Pangnirtung, from 1967 through 2002 (McElroy 1973, 1977, 1996) and Qikiqtarjuaq (or Broughton Island, in English) and Cape Dorset (or Kinngait) in 2002 (see Figure 5.1). In each settlement I boarded with Inuit families and did participant-observation. With the use of interpreters, I conducted the interviews in 1999 and 2002 at elder centers, in people's homes, and at an assisted living center.

Thirteen people were interviewed in 1999 and 30 in 2002, with a total sample of 43, including 25 women and 18 men. The sample ranged in age from 55 to 88. Nine women and five men (33% of the sample) were under 64 years old. The average age of the women was 67, and the median age was 73. The average age of the men was 74; the median age was 73.

The mayors, senior administrative officer, and town councils of the four communities reviewed both English and Inuktitut versions of the research proposal, and each community approved the research. Informed consent statements, translated into Inuktitut syllabics, were presented to each participant, most of whom read syllabics. Interpreters read the consent information to the few who could not read the syllabic script. Participants received honoraria, and interpreters were paid by the hour. Offered the choice of using pseudonyms or their real names in published material, all insisted that their real names be used.

The largest of the four towns in this project is Iqaluit, meaning "fish." Originally called Frobisher Bay, Iqaluit expanded during World War II with construction of an American air force base for refueling and maintenance of fighter aircraft going from the United States and Canada to England. After the war, the United States continued to use the base as a transshipment point for materials going to the Thule airbase in Greenland (MacBain 1970). In 1959, Iqaluit became an administrative center with a native population of about three hundred. By 1963, the town had grown to nine hundred Inuit and seven hundred nonnatives from southern Canada and other Commonwealth countries, and from that point it grew into a miniature metropolis to which jet aircraft transported government employees, construction workers, doctors, teachers, patients, students, and tourists. All children attended school, with English being the language of instruction. In sum, 87% of Inuit men and 31% of Inuit women were wage employed. The average Inuit household size was 6.5 people; housing was subsidized; taxes were minimal; and medical care was free.

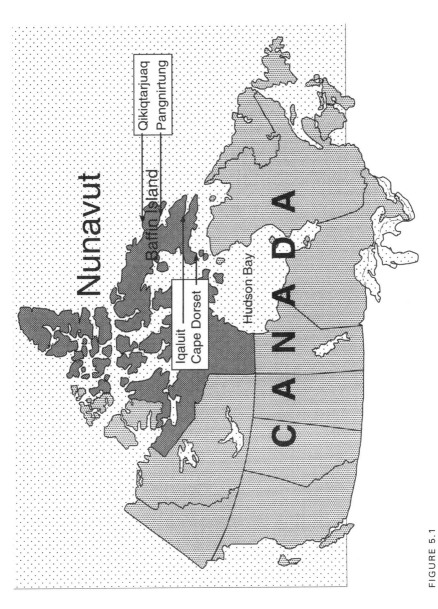

FIGURE 5.1
Map of Nunavut, showing locations of four research sites.

Cape Dorset, a second field site, is located on Dorset Island, on the south shore of Foxe Peninsula, jutting into Hudson Strait. Inuit traded with whalers who came into the area in the 19th and early-20th centuries, and they also enjoyed trade and social interaction (as well as occasional competition) with people living on the shores of arctic Quebec. In 1913, a Hudson's Bay Company trading post was built at Cape Dorset, and the Baffin Trading Company opened a post in the 1940s. Two churches (Catholic and Anglican) and a school were built there in the early 20th century. In 1953 James and Alma Houston began art programs that became the West Baffin Eskimo Cooperative, supporting indigenous artists and marketing prints and carvings throughout the world. In the 1950s people began to move into town, and new houses were provided, but as late as 1968 Cape Dorset remained a small settlement, with many families still oriented to hunting. Since then, Cape Dorset has become a relatively affluent community of about 12 hundred people, about 90% Inuit. Many affluent visitors arrive in private jets and represent art galleries and private collectors. A number of carvers and printmakers receive large commissions and steady royalties for their work. Employment rates are generally good in Cape Dorset, with full-time jobs in government, construction, education and health, language interpretation, and tourism.

The third site is Pangnirtung, where I began research in 1969. Pangnirtung was a trading post and mission established in 1921 on Pangnirtung Fiord on the north side of Cumberland Sound. Some families worked for the police or traders and settled there, but most of the population remained dispersed in small hunting camps. The RCMP and missionaries traveled by dog team to these camps to maintain a census and to provide services. Inuit periodically came to Pangnirtung to trade, seek medical care, and celebrate Christmas.

In the 1950s, the federal Department of Health and Welfare posted medical officers to the mission hospital for two-year terms at Pangnirtung. Their duties included medical patrols with the RCMP and with the interpreter–guides. Canine distemper swept through the Cumberland Sound camps in 1962, killing many dogs and stranding hundreds of people, who had to be evacuated by plane to Pangnirtung, where temporary shelter was provided. Some returned to the land, but many remained in town (Haller 1966). By 1970, about 600 Inuit and 50 Euro-Canadians lived in Pangnirtung. Most native families occupied prefabricated houses with electricity and phone service but no plumbing.[1] Few families had been able to replace their dog teams, and many purchased snowmobiles instead. Of Inuit men in Pangnirtung, 20% were wage employed, and 40% of all households had one or more persons employed. Many of the men continued to hunt, and women processed skins for household use and for trade. In the 1970s, sealskins were profitable, often bringing CAN$25 or more at the Hudson's Bay Company store. Young women were trained in weaving at the crafts center, over time becoming respected professional weavers.

Qikiqtarjuaq ("big island"), the smallest of the four communities in this project, is situated on an island on the northeast coast of Baffin Island facing Davis Strait, 96 kilometers north of the Arctic Circle. John Davis explored the area in the late 16th century, and Robert Bylot and William Baffin explored in the 17th century. Both commercial whaling and fishing were well established in the area by the 19th century. Few families lived at Qikiqtarjuaq until the 1950s. Earlier, people congregated at whaling stations seeking employment, such as at Kivitoo to the northwest and at Padloping Island 90 kilometers south, where there also was a military weather station in World War II. In 1955, a Distant Early Warning Line site constructed at Qikiqtarjuaq attracted families seeking employment. People continued to migrate seasonally between Kivitoo and Padloping, as well as to Clyde River to the north and to Pangnirtung through the mountains to the east (or by boat through Cumberland Sound). In the late 1960s families were pressured by the federal government to settle at Broughton Island.

Qikiqtarjuaq has around five hundred residents, 95% of whom are Inuit. Various branches of the territorial and federal governments employ many adults. The Northern Store (formerly the Hudson's Bay Company store) also offers employment. There is little private enterprise in this small community, although outfitting companies provide transportation and equipment to hikers, climbers, tourists, and scientists. Crafts, including skin work, carving, and knitting, provide some income. Well-known local musicians and songwriters perform at various events and festivals throughout Canada and are successfully marketing their CDs.

Inuit history is indeed rich and multifaceted. In the following, I present data from interviews with Inuit elders that present this history from a local perspective. As a whole, the memories presented illustrate the links among subsistence, mobility, and integration into national and international arenas of indigenous activism.

MEMORIES OF THE PAST

They killed a whale the day I was born. They were thankful when I was born, as it was the end of the whaling period and there weren't many whales left. Sowdlu, a 76-year-old seamstress and midwife, recalled that her birth in 1923 was a day of celebration for her family and others in her hunting camp, as a whale had been taken that would provide food for weeks and oil to be rendered and sold at the Hudson's Bay post. Most of the 43 elders interviewed were born in out-post camps and lived in traditional hunting camps as children. Henry, age 77, remembered, *My very first time I caught a caribou was in the Igloolik area. I was very young at the time. I think I was taught to hunt my first caribou.* He added, modestly joking about his accomplishment, *I believe it was a sickly caribou!*

Three of the interviewees grew up in small settlements, their parents employed by the RCMP, the hospital, or by the Hudson's Bay Company. Few had formal education, although most learned to read and write syllabics from Anglican prayer books. With few exceptions, all learned traditional, gender-specific skills from parents and grandparents through observational learning rather than formal tutelage.

At least a third of the respondents suffered deaths of parents or siblings in childhood or adolescence. The surviving parent might have remarried, or another family adopted the child. Memories of the stepparent or adoptive parents were sometimes positive, but there were also stories of children made to work very hard by stepparents or adoptive parents. Whether with natural or adoptive parents, teenagers were expected to contribute to subsistence tasks or find employment in the settlements. Evie, age 75, recalled, *I came to work here at the hospital when I was 13 years old. I did dishes and mopped the floor. And my duties increased as I got older, and I got promoted to doing the laundry. I worked there until I was 20 and got married.*

As respondents were growing up in the 1920s and 1930s, British and Euro-Canadians were already influencing their families' lives. Many remembered the booming voices of Anglican ministers. Some said they were frightened of white people, whom they called "Qallunaat" (after the Inuktitut term *qallu*, meaning "eyebrow," perhaps because many of the newcomers had bushy eyebrows), and they recalled vividly that store clerks had unpleasant body odors of tobacco, kerosene, and soap. One man humorously described how people would come back to the camps smelling like Qallunaat after trading in town.

Recalling her childhood in the North, 63-year-old Annie liked to remember the attention she received when she danced for southern Canadians. *I remember clearly, the Mounties and the minister used to come once a year to our camp. We would gather in one place, and my mother would play the accordion. I would dance for the Mounties. They always asked me to dance because I would dance from the heart.*

Some remembered Qallunaat as kind and generous, but there were also examples of police, doctors, and traders asserting authority over Inuit's decisions where to live and when to receive medical care. Sometimes these decisions benefited Inuit, but compulsory moves into settlements and forced hospitalizations and schooling left residues of resentment and ambivalence. Jacopie, age 65, recalled, *For three years the government was working toward relocating all of us. Three times they told us to go, but we didn't want to, so three times we said no. But then they were forcing people, and we finally said yes. The main purpose the community demanded that people from Padloping move up here was so they could get houses. They said that more people would be dying down there at Padloping, and if they got sick, the police would not help them down there. They would just be dying. It was quite stressful for those people, especially for the elders.*

Mobility was a fact of life for people in hunting camps and for those engaged in seasonal employment. Memories of childhood reflect frequent moves over wide areas. For example, people living in Qikiqtarjuaq in 2002 had traveled in childhood and earlier adulthood throughout the east-central Baffin Island region. They traveled among camps and settlements to trade and seek medical care, visit relatives, find spouses, and search for employment. Some went as far as Ontario or Quebec to be hospitalized for tuberculosis. But voluntary mobility of individuals and nuclear families differed from forced relocations of entire camps, which many elders remembered with bitterness.

Critical histories of the North have emphasized that most Inuit were compelled to move into settlements for administrative purposes, but the life stories in this project show varied reasons for moving into town: illness and need for medical care, seeking employment or joining a spouse or family member who had found employment, and government pressure to move. Three moved into town as adults so that they could be near hospitals or nursing stations for management of their health problems, and four came as the children or young spouses of persons needing to relocate due to illness or disability. Ten cited employment as the reason for moving. Three moved as children with parents seeking work, and three moved with spouses who were seeking work or had already found employment at the air force base in Iqaluit, on the Distant Early Warning Line, or elsewhere. Four men came into town seeking work so that they could support their large families who were left behind in camps but eventually relocated.

Government pressure was the reason given by nine respondents for moving. These were moves of desperation that followed famine, epidemics, and the deaths of sled dogs. Some of these families hoped to return to their camps, but pressure on them to keep their children in school influenced decisions to settle in or near town. Three relocations were related to marriage opportunities. Two women described marriages that had been arranged for them in another settlement, and one widower went to a community specifically seeking a wife. Akakaq, age 81, recalled, *I decided to move here on my own; no family, no wife with me. I was hoping I would find a girl when I got here. The reason I moved here is that I had a wife and she passed away, and the two children passed away. It's really hard to live on the land without a partner.*

Two individuals were adopted by families in other communities after having lost one or both parents to illness or accidental deaths. The remaining cases involved a series of moves between camps and towns over several decades of married life (including second marriages) and then finally settling in town in middle age. A major motive for moving back and forth was to be near relatives and in-laws.

What did respondents remember about settling in town? One man recalls being "awestruck by all the noise" of construction, motor vehicles, snowplows, and airplanes. Used to the quiet of living on the land, he found town life to be stressful. Some remembered feeling "shy" when they moved into town, like they were "sticking out," conspicuous. Some felt homesick and lonely for relatives. One woman expressed longing for her home region: *I needed to get a good breath. I would stretch out my arms and breathe deeply to try to feel better.*

The mix of people from different regions was a source of tension. People described differences in dialect, hunting methods, dress, use of alcohol, religious practices, and personal mannerisms as problems in the early days of the settlements. Some mentioned positive aspects, such as pleasure of getting houses to live in and finding a job, as well as gratitude for health care for themselves and their children. Those with disabilities found life easier. Many remembered specific kindnesses and assistance shown to them during emergencies. Speaking about treatment for a hand injury as a young man, Oqutaq, age 66, recalled, *I was fortunate to go back to my camp by airplane. The nurse here was very fond of our baby, and he was worried about me. So they went to my camp by plane. It was overwhelming how we were cared for at that time when I needed medical attention.*

One aspect of Inuit–European interactions is the extent of unfairness in labor relations. At least a third of the respondents described working for little or no pay, other than some food and clothing, or exchanging skins for supplies instead of money. For many, no money changed hands to compensate for years of labor. In some cases, employees were still in their early teens, faced with heavy physical work involving laundry, cleaning, cooking, and provisioning dog teams in return for food and clothing and sometimes shelter. Evie, age 75, remembered, *I didn't get paid for the work, but they provided clothing and food and a place to stay. I never got money, but after I got married, I got a substantial amount. Apparently the nurses had been saving my money. The whole time I was working there, for seven years, I never got paid, but they had been saving money for me, and that was very convenient when I got married. I don't remember my father getting any of my pay, but sometimes he did get bullets from the hospital, so I figure I was responsible for that.*

Some respondents expressed bitter memories of being forcibly separated from their families for schooling or medical reasons. Others spoke of family members who were taken South, who died in the hospital, and were buried in unmarked graves in Quebec or Montreal. Although 50 or 60 years had passed, the pain and trauma of these losses were still evident.

Probably most bitter are the accounts of relocation. For example, Henry remembered, *And then I went to Hamilton [Ontario] due to illness. It was not TB, but a problem in my stomach because of a hunting injury. And then the government asked me to move here a number of times after that, so I decided to say yes. I really wanted to return to the community where I lived [Pond Inlet], but the people I was working for kept telling me I couldn't go back. I had to live in a warm place because of my injuries, they kept telling me. I couldn't go back, because I wouldn't have a house if I go back. My wife had passed away by then, so I had no choice.*

Administrators may have failed to recognize the deep attachment that people held to their familiar hunting areas and campsites. Perhaps they thought that these feelings would fade. However, incidents from the 1950s and 1960s remained vivid in respondents' narratives many years later, adding echoes of regret and resentment to their life histories. Those who lost their connection to the land and thereby lost part of their identity were especially poignant. As Pitalusa related, *I had to go down South for medical reasons. I was eight years old. I came back home when I was 15 years old. As a young woman you are expected to know what to do with skins in winter, how to sew and make clothing, but I didn't know any of that. And yet I was expected to. It was such a hardship for me.*

HEALTH AND DIETARY CHANGE

"Country" (local) food, *niqissat* in Inuktitut, is an integral part of traditional Inuit culture. Access is largely based on the seasonal availability of marine and terrestrial mammals, particularly seals of various species, whales, walrus, and caribou, as well as fish and shellfish. Small game includes ducks, geese, ptarmigan, other fowl and their eggs, and arctic hare. Some mammals are hunted or trapped but not usually consumed, including fox, wolves, and polar bear. In the past muskox were hunted and consumed, but they are now a protected species. Plant life has always played a minor role in the diet, although berries, sorrel, and kelp are gathered seasonally. Plants and animal skins were used for medical purposes in the past.

For the last 40 years, eastern arctic Inuit have adopted a mixed diet. Even the smallest settlements have plentiful supplies of basic staples: flour, sugar, pasta, rice, tea, oatmeal, biscuits, and canned vegetables. Fresh fruit, vegetables, and dairy products have been seasonally available at high cost for several decades and are now flown in weekly. Stores in larger towns rely on daily air traffic for deliveries of produce, dairy, meat, baked goods, and staples.

Most Inuit consume some country food daily, especially fish, seal, whale skin (*mattaq*), and caribou, and some men continue as full-time hunters. The demand for land food has meant that many wage-employed men hunt on weekends and entire families travel by boat or snowmobile to hunting camps during winter and summer holidays. Fishing and seasonal harvesting of long-necked clams, other mollusks, kelp, duck eggs, and berries supplement the diet.

Food-sharing networks are an essential part of the social structure in all of the communities. When people travel from one settlement to another by plane, they often take frozen fish or caribou to give to those whom they are visiting. In smaller towns such as Pangnirtung and Cape Dorset, land food is far more available than in Iqaluit, and thus distribution between settlements is an important part of access to traditional foods.

A typical day's diet for a 50-year-old employed man in Iqaluit is homemade bannock (an unsweetened scone) and tea in the morning; barbecued ribs, fries, and a coffee or tea from a fast-food place for lunch; a snack of *mattaq* dipped in soy sauce with tea after work; and freshly killed, often uncooked seal or caribou at a relative's house later in the evening, sometimes accompanied by meat broth, mashed potatoes, bannock, and tea. The diet is low in dairy and fresh fruit, although vitamin C is plentiful in whale skin and in raw seal. Some adults consume commercial food at restaurants around three or four times a week and confine their consumption at home to bannock, tea, and wild game (mostly seal), but children have commercial food every day except when they are in hunting camps.[2]

Food security is not only a nutritional issue but also one of cultural integrity. Many Inuit believe that seal meat is essential for health and strength, a "rejuvenator of human blood" and a "life-giving" nutrient (Borré 1991:54). Despite the inroads of commercial food on everyday meals, country food is considered more healthful than imported foods and is preferred at community feasts, weddings, and funerals; on holidays; at hunting camps; and when visitors arrive from other settlements. One man in his 60s pointed out, *White man's food makes you weak because it's white, like white bread, white noodles, white rice. Our food is dark red, like blood. It makes you strong. Our children grow better on our own food.* M. M. R. Freeman mentions, "It is also recognized that old people have a special 'need' for meat and that eating meat fresh and raw is necessary to maintain body warmth during cold weather" (1996:59). In recent years, a central component in social services for elders is to purchase meat from hunters and thereby ensure availability of freshly killed game for consumption at the elder centers, in assisted living apartments, or at home. When I asked about food access, Josea, age 61, answered, *Yes, I get enough country food. I always eat it, because it's the only thing that stays in your stomach for a while. You can feel it in your stomach for a while.*

The sharing of land food also symbolizes positive aspects of cultural identity, as Kopa, a woman of 62, emphasized: *My son usually provides country food for me, but he has a job and has to travel, and he's not here now. The thing about Inuit is that everybody is so supportive of each other, and even if I don't have anything at the time, and even if I'm not asking for it, people will come and give me country food like caribou, mattaq, fish, or seal. Country food is my main source of food because I grew up with it. If I don't have it for too long, I start craving it really bad, and I need it most of the time.*

In 1999 I asked respondents about their sources of country food and whether they were satisfied with the amounts they received. Seven of the 13 respondents said that while relatives and neighbors were generous in sharing food, they did not get enough land food to eat daily, as they preferred to do. In 2002, only 5 of the 30 respondents said that they did not get enough country food; 25 affirmed that they were getting adequate country food provided by their own subsistence activities, by their children and other relatives, or as purchased frozen food. The main complaint from the active hunters was that the cost of gasoline prevented them from hunting as often as they wanted. No elders mentioned being aware of PCBs (polychlorinated biphenyls) or other chemicals in land food or the need to limit intake of marine mammals, and even though global warming was being discussed on television and radio shows, no elder brought up this topic in interviews.

Do elders expect continued access to land food? Some worry that the numbers of seal and beluga are declining, and some mentioned the rise in gasoline prices as a constraint. Others, mindful of the cost of imported food, believe it is essential to maintain traditional subsistence. There is uncertainty about economic stability under Nunavut. Jayko, age 77, said, *It will come to a point where it will be impossible to live off the economy the way it is right now. It will be almost impossible to buy food with money. I think that will happen one day, so we cannot forget the Inuit ways, because we can live off the land, and that way we will be able to survive.*

Freeman notes the serious need in the arctic to safeguard food resources, "which are threatened from two main directions: (1) the potential of localized over-exploitation as human populations grow in size, and (2) environmental damage" (1996:65). Environmental threats come from oil and natural gas exploration, mining, runoff of toxic wastes into rivers and bays (and ultimately to ocean waters), sewage plants, construction, airfields, and so on. As early as the 1980s, it was recognized that the tissues of seals, whales, and caribou contained PCBs, of which 20% of the Qikiqtarjuaq people had consumed more than the acceptable level. Among children, two-thirds had higher-than-acceptable levels of PCBs in blood samples (Kinloch and Kuhnlein 1988:160). Inuit infants exposed to persistent organic pollutants through breast milk had higher rates of ear infections than did bottle-fed infants (Dewailly et al. 2000).

Greater connection with the national and international communities has brought an array of health changes in Baffin Island communities. On the negative side, high rates of smoking and alcohol use among adults and teens, lower nutritional quality in commercial food, and a more sedentary lifestyle are contributing to increased rates of respiratory and heart disease, diabetes, and cancer. On the positive side, immunizations and good medical care have decreased child mortality and increased the average life span. Tuberculosis, which affected most of the interviewees in childhood and the young adult years, has been contained. Those who are aware of toxins in marine mammals have concern about long-range effects of contaminants on health, but at this point there is greater public health focus on reducing smoking and alcohol use than on restricting consumption of land foods. Yet, for many Inuit, access to land foods is essential. Pitalusa, a 60-year-old woman, explained, *We have to eat. And for most of us, we have to have our traditional food. I'm a diabetic, and I'm allowed certain foods in the Qallunaat way, but I find eating my traditional food makes my blood better and helps me. Every so often I'm asked if I have any ancestors with diabetes, and I say I don't know. I don't know that kind of sickness from the past.*

All but two of the research participants were ambulatory and in relatively good health. One lived in a shared bedroom in a facilitated living residence, where she received meals and daily nursing care for chronic obstructive pulmonary disease, which affects many elderly Inuit. Another person with respiratory problems lived in a facilitated living apartment adjacent to the nursing facility. A woman with poor vision depended on her family for daily assistance.

Most of the interviewees described problems with arthritis, especially knee problems. Eight had undergone knee and hip-replacement surgeries. One woman had a hip deformity present since childhood partially corrected in her 60s, and another had reconstructive jaw surgery. Three women and one man had type 2 diabetes. Three respondents had received pacemakers and shunts to correct arterial blockage and fibrillation, and five used inhalers for asthma.

Despite these health complaints, most respondents had active lives. Men through their 70s continued to hunt, ride snowmobiles or four-wheelers, fish, carve, and take part in church activities. Women continued participating in sewing and crafts, food preparation, child care, and church activities. Although bus and van transport was available to bring elders to activities and for medical appointments, a fair number preferred to walk to community events, to the stores, and to visit friends, especially on warm days. One man in his 80s used a four-wheel ATV (all-terrain vehicle), complete with helmet, to get around the settlement. In summer and fall, elders accompanied family members on trips to the land to dig for clams, gather berries, participate in caribou hunting, and share campsites with close relatives. Family reunions were held at ancestral campsites, giving people a chance to honor ancestors buried at these sites and other deceased family members.

REFLECTIONS ON NUNAVUT

Throughout the 1970s and 1980s, organizations such as Inuit Tapirisat of Canada pushed for land claims negotiations and mobilized interest in increased self-governance (McElroy 1980). Publication of the *Report of the Inuit Land Use and Occupancy Project* (Freeman 1976) documented historic and current indigenous land use patterns, information essential for negotiations. A proposal to consider a new territory to be called Nunavut was presented to the federal government in 1976. In 1982 Inuit Tapirisat created the Tungavik Federation of Nunavut to represent the 26 communities in settling land claims. Studies of land use and resources, surveying three hundred million hectares, were carried out with the participation of the communities (Riewe 1991). An agreement in principle was reached in 1990.

In 1992 voters ratified the Nunavut agreement in principle, an important step toward the final establishment of the territory seven years later. Key elements in the agreement included (1) Nunavut territory to be officially established April 1, 1999; (2) title to 355,842 square kilometers of land to be ceded to native residents of Nunavut; (3) mineral rights to a portion of that area to be ceded; (4) compensation of CAN$1.1 billion to be paid to Inuit and to Inuit-run corporations over 14 years, beginning in 1993; and (5) surrender of Aboriginal title so that general claims cannot be pursued in the future (Crowe 1999).

In the early years of Nunavut governance, the prospect of legislation affecting hunting and harvesting rights of indigenous people has been central in public discourse and debate. Although hopeful that the government will ease quotas and allow some hunting of protected species, many residents fear the influence of environmental activists. International and territorial policies influenced by this activism may continue to threaten the health ecology of Nunavut residents.

Nunavut has created a construction boom and upgrade of basic infrastructure unparalleled in Baffin Island history. Iqaluit, with about five thousand residents, has become the territorial capitol and faces a severe housing shortage. Handicapped-access housing and assisted living facilities for elders and people with disabilities have been constructed, along with community centers for elders.[3] The town offers cosmopolitan amenities—gourmet coffee-houses, a local Internet provider, hockey rinks, libraries, televised bingo on a local channel. In summer the town is overrun with tourists, trekkers, bureaucrats, journalists, and scientists.

Traffic jams are a common sight, especially at midday. Construction is everywhere, and houses are built on precarious sites, suspended over deep gullies and supported by tall piles drilled deep into permafrost and bedrock. As yet lacking a system of street names, in 2002 the town held a series of meetings with local residents to try to reach consensus on names that would honor Inuit leaders and recognize traditional place-names. By successfully hosting the Arctic Winter Games in March 2002, the town gained new sports facilities and built its reputation as an attractive tourist and conference location.

Pangnirtung has remained a more compactly settled and traditional community, with approximately 12 hundred residents (90% Inuit). Commercial scallop fishing provides employment, and tourism (including cruise ships) has become a major industry. The beauty of the fiord, surrounded by massive mountains, is unequaled, although the town is equally famous for high winds and unpredictable weather. Many trekkers pass through Pangnirtung on their way to Auyuittuq National Park Reserve, a 21,500-square-kilometer area with challenging trails through glacial mountains and river valleys. Travelers visit Kekerten Historic Park to see remains of a whaling station. Pangnirtung itself features the Angmarlik Centre (an interpretive center), a weaving and print-making shop, a summer music festival, and an Anglican clergy training school. New housing units, including apartments designed for elders and people with disabilities, were built in Pangnirtung between 1994 and 2002, but there is still a housing shortage, and land for new construction is scarce.

The long waiting lists for housing are evidence of uncontrolled population growth in Qikiqtarjuaq and Cape Dorset as well. People speak of increasing drinking and violence in the settlements and express dismay at rising prices of food and clothing. Inuit in all three of the smaller settlements perceived Iqaluit to be a much more crowded and violent community, with greater problems of life quality—housing shortages, pollution, noise, high prices, dangerous roads, a lifestyle that is too hectic and pressured where there is less sharing of food among neighbors and kin and where people are getting fat from eating fried food.

In 1999, when asked about their opinions of Nunavut, some elders expressed a "wait and see" attitude about the government. One man said, *I really have no opinion about it because I can't see it, or smell it, or touch it. It is not solid. So I don't have anything to say about it.* Some said they liked the fact that elders would be the first to receive compensation checks. Others were critical or ambivalent, saying that elders were not being consulted enough and changes were not coming quickly enough.

By 2002, the new government had initiated many changes. Compensation payments were being dispersed. Identity-building efforts were evident in the media, in community events and ceremonies, and in the rhetoric of politicians. I was curious to see whether elders' attitudes about Nunavut had solidified and whether they identified with the new territory.

The 30 elder Inuit I interviewed in 2002 had mixed feelings about the new territory. Most expressed pride in Nunavut, but there was also a sense of frustration with continued hunting regulations, quotas on specific animals, and new laws restricting use of rifles by youths. Some felt that elders should be consulted more, so that traditional values, such as the importance of children's learning to hunt, would be maintained. The strongest sentiment was dissatisfaction with the slow pace of decision making by representatives. Jamesie, age 73, said, *The people who are involved with Nunavut, they don't ask the elders for their input. We would like to be questioned, asked what we think of Nunavut. It's fine that we have Nunavut, but there are so many restrictions on wildlife and quotas. Whenever you kill an animal after the quota has been filled, they can charge you. We live in such a huge area, and people hunting in this area will not know that these hunters had already filled the quota.*

Amidst this change in governance, regional identity appears to have persevered. One woman in Iqaluit, for example, gave the following opinion: *Yes, I think definitely my family and I are Nunavummiut. We are from Nunavut. There was no identity for the Inuit before. Different countries got their identity from where they were. After Nunavut, we have been able to call Inuit "Nunavummiut," and I think that's a good idea, because they had no other name than "Eskimo."* Another woman said, *I think Nunavut is getting better. They are still in the planning stages of development. I really think it's going to get better after all these plans are in effect.* But when asked whether she regarded herself as Nunavummiut, she said, *No, I'm Kanatamiut [Canadian].*

My question of whether residents of the four communities had distinctive identities was a difficult concept for respondents to grasp. Several responded by saying, *We all are the same in God's eyes. Even Inuit and Qallunaat, they are the same, no important difference.* This kind of answer illustrates the egalitarian ethos of traditional Inuit culture. Most frequently mentioned were dialect differences as a way of identifying a person's region of origin. Physical differences were generally not mentioned, although some noted that people in Pangnirtung had European features, including light eyes and hair. Some mentioned negative stereotypes of the towns themselves—sometimes mentioned in jest, sometimes more seriously. Iqaluit is viewed as being too noisy, too crowded and competitive. Qikiqtarjuaq is considered too small and isolated; one young man in Pangnirtung turned down a job offer there because he said there was "nothing to do" in such a small town. People in Iqaluit expressed the perception of Cape Dorset as a dangerous, violent place with a long history of having feuded with people from northern Quebec.

Stevenson's 1997 genealogies from the early 20th century document the frequency of marriages between people of the Cumberland Sound region and those of the east coast (Kivitoo, Padloping, and other camps). Life histories in this project suggest that the people of Pangnirtung and Qikiqtarjuaq formed a regional unit bound through intermarriage, exchanges and visiting, and friendships. Conversely, the people of Iqaluit are more closely tied to those from Cape Dorset and in Kimmirut (Lake Harbour), forming a second regional unit with ties based on marriage, adoption, and migration histories. Although the distances by land are not far, just a few hundred kilometers, the social distance between people is considerably greater.

Cumberland Sound Inuit attach great importance to locality in their sense of identity. Kinship plays a major role in determining local group composition, but "individuals are usually regarded as members of a specific local group if they reside permanently with that *nunatakatigit*, regardless of their birthplace or primary kinship connections" (Stevenson 1997:193). Thus marriage ties people together almost as strongly as biological kinship does and probably connects people across geographic boundaries as well. Mark Nuttall's research on environment and identity in Greenland indicates similar patterns: people use memory as a way "of articulating the relationship between community and landscape . . . in a fusion of cognitive and spatial symmetry.

The environment is perceived in a particular way and people are involved in a dialogue with a landscape suffused with memory and highly charged with human energy" (1992:57).

In the Baffin Island study, there was variability in local versus territorial or national identity. Some respondents identify with being Nunavummiut, but many identify themselves more as Canadians, Inuit, or residents of their towns or regions. The Nunavummiut identity is being promoted by the media, but such a fundamental change may not be compatible with regional identities that have persisted based on social and marital ties, internal migration, and common dialects.

A VISION FOR NUNAVUT

Hunting and other traditional practices, and their relationship to natural resources, are integral components of Inuit culture and are perceived as being vital to maintaining independence. Henry, age 77, emphasized that natural resources must be preserved for future security: *Inuit do have two hands. Even if we're not office people, we still have practical skills. These mining operations do provide income, but they shut down. I believe that hunting and sewing are the major source of income that can go on and on. The skins and furs can be used for crafts and for arts. The resource is always there if it's used wisely. . . . It would be more of a balance if Inuit have a way of making money, then the economy would balance. We wouldn't be so dependent on the stores. It wouldn't be so one-sided.*

Many of the people interviewed held a vision of a future in which Inuit culture and language would persist, a culture not merely of carvings in art galleries and artifacts in museums but of people procuring food and materials for their livelihood from the oceans, rivers, and mountains. It is a culture in which resources are distributed and shared, waste and excess are avoided, and egalitarian rules of conduct continue. Elders are a rare repository of ethnohistory, with memories of hardship and unfair treatment as well as endurance and personal resilience. Their stories have given us a glimpse into an important chapter in the history of the North, and their hopes for the future give a sense that the culture will persist.

Over the past century the human niche in arctic ecosystems has been transformed through successive stages of culture contact. Once nomadic hunters, indigenous Baffin Islanders live today in modern towns and hamlets, working as wage earners. Their children attend schools in which cultural inclusion programs help to prepare the next generation for new roles and skills while re-taining identification with the traditional culture. Families enjoy new forms of media and recreation, good health services, and facilities for elders and for people with disabilities that allow them to remain in their home settlements.

Throughout all this change, Inuit identity remains strong. Much has been written about Nunavut as an experiment for the future, but what is over-looked is how the residents of Nunavut interpret the recent past, not the ide-alized aboriginal past, but rather the experiences of recent years. In consultation with Inuit interviewees and interpreters, I have attempted to identify in this study the ethnohistorical and ecological models of change held by Inuit, as those expressed in their life stories and opinions about political and environmental change and access to traditional foods.

Taking a long-term view of the spectrum of transformations in four com-munities of south and central Baffin Island, I have used a model of health ecology that may apply to other indigenous communities striving for in-creased political autonomy. Among the central variables in this model are (1) access to traditional foods, not only for nutrition, but also for stability in food-sharing networks; (2) transgenerational effects of unresolved losses, dis-locations, and exploitation by colonial agents; and (3) persisting regional and local identities connected to community networks and to a geography of meaningful spaces and relationships. Each variable is an integral component of the well-being of Nunavut's citizens.

As early as 1965, John and Irma Honigmann noted the national and global connections feeding into the maintenance and provisioning of Baffin Island communities. They wrote,

> Many valuable services would not be forthcoming except for the many-stranded lifeline that links Frobisher Bay to the outside society of Canada and thence to the rest of the world. In that wider society originate most of the satisfiers that make the town a comfortable place to live ... [but we] note that the Eskimos have lost as well as gained something in achieving the tremendous diversity of satis-factions, ranging from outboard engines to medicines and chewing gum ... [as] they have had to shed a measure of their former autonomy. [1965:53]

Forty years and two generations later, the Honigmanns' description of the "lifeline" between the region that would become Nunavut and the rest of the world remains valid. This many-stranded linkage now depends not only on the exchange of goods but also on transmission of information via the Internet, electronic mail, satellite television, distance learning, and teleconferences. Inevitably, that information includes news of environmental threat, health risks of unchecked contamination of marine waters, and the potential disruption of global warming. Heeding those warnings, the leaders of Nunavut are primed to use legislation and regulations designed to preserve their ecology as much as possible. Although some sources of environmental degradation are beyond their direct control, mobilization of international opinion against the major polluters offers the best chance for eventual habitat protection and restoration so essential for continued health and well-being in Inuit communities.

NOTES

I gratefully acknowledge support of the 1992 research by the Canada–U.S. Trade Centre in Buffalo, New York; support of the 1994 research by the Canadian Studies Grant Programs, Faculty Enrichment Programme; and support by the Canadian Studies Research Grant Programme, Office of Academic Relations, for the 1999 project "Ethnohistory of Political and Ecological Change on Southern Baffin Island" and the 2002 project "Microcultural Diversity in Ethnic Identity: Inuit Elders' Narratives of Change in Four Nunavut Communities." The project would not have been possible without the excellent assistance of the interpreters: Leetia Papatsie in Iqaluit, Andrew Dialla in Pangnirtung, Sarah Kuniliusie and Harry Alookie in Qikiqtarjuaq, and Aksatungua Ashoona and Akalayuk Kavavow in Cape Dorset.

1. The homes of teachers, administrators, and police usually had plumbing and hot water. This disparity was not resolved until the 1990s, after new housing was built.

2. There is considerable variation. I boarded with a young couple that only ate store food at home, principally soup, sandwiches, canned stew, pizza, and eggs, but who had at least one meal of land food every day at their parents' homes.

3. The Iqaluit Elders' Facility offers a day program from noon to 4:00 P.M. weekdays, where elders and visitors may obtain a lunch of country food (e.g., seal, caribou), soup, biscuits (*palowak* in Inuktitut), canned fruit, dessert, and tea for about 50 cents. Pangnirtung has an elders' room at the Angmarlik Centre where people play cards, talk with visitors, sew and knit, and have tea and snacks. Qikiqtarjuaq and Cape Dorset have elders' committees but no separate facilities.

REFERENCES

Boas, Franz
1888   The Central Eskimo. Annual Report, 6. Washington, DC: Bureau of
        Ethnology, Smithsonian Institution.

Borré, Kristin
1991   Seal Blood, Inuit Blood, and Diet: A Biocultural Model of Physiology and
        Cultural Identity. Medical Anthropology Quarterly 5:48–62.

Chance, Norman
1990   The Iñupiat and Arctic Alaska. Chicago: Holt, Rinehart and Winston.

Crowe, Keith
1999   The Road to Nunavut. In Nunavut '99. M. Soubliere and G. Coleman, eds.
        Pp. 24–39. Iqaluit, NWT: Nortext Multimedia.

Dewailly, E., P. Ayotte, S. Bruneau, S. Gingras, M. Belles-Isles, and R. Roy
2000   Susceptibility to Infections and Immune Status in Inuit Infants Exposed to
        Organochlorines. Environmental Health Perspectives 108(3):205–211.

Dickerson, Mark O.
1992   Whose North? Political Change, Political Development, and Self
        Government in the Northwest Territories. Vancouver: University of British
        Columbia Press and Arctic Institute of North America.

Eber, Dorothy H.
1989   When the Whalers Were Up North: Inuit Memories from the Eastern
        Arctic. Montreal: McGill-Queen's University Press.

Freeman, M. M. R.
1996   Identity, Health, and Social Order: Inuit Dietary Traditions in a Changing
        World. In Human Ecology and Health. Maj-Lis Foller and Lars O. Hansson,
        eds. Pp. 57–72. Goteborg, Sweden: Goteborg University, Section of
        Human Ecology.

Freeman, M. M. R., ed.
1976   Inuit Land Use and Occupancy Project. Ottawa: Department of Indian and
        Northern Affairs.

Government of Nunavut
2004   Consensus Government. Electronic document,
        www.gov.nu.ca/Nunavut/English/about/cg.pdf, accessed October 2004.

Hall, Charles F.
1970   Life with the Esquimaux. Rutland, VT: Charles E. Tuttle.

Haller, A. A.
1966   Baffin Island–East Coast: An Area Economic Survey. Ottawa: Department
        of Indian Affairs and Northern Development.

Hankins, Gerald W.
2000   Sunrise over Pangnirtung: The Story of Otto Schaefer, M.D. Calgary: Arctic
        Institute of North America.

Honigmann, John J., and Irma Honigmann
1965   Eskimo Townsmen. Ottawa: Canadian Research Centre for Anthropology,
        St. Paul University.

Kinloch, D., and H. Kuhnlein
1988   Assessment of PCBs in Arctic Foods and Diets—a Pilot Study in
        Broughton Island, Northwest Territories (NWT), Canada. Arctic Medical
        Research 47(Suppl. 1):159–162.

McBain, S. K.
1970   The Evolution of Frobisher Bay as a Major Settlement in the
        Canadian Eastern Arctic. M.A. thesis, Department of Geography,
        McGill University, Montreal.

McElroy, Ann
1973   Modernization and Cultural Identity: Baffin Island Inuit Strategies of
        Adaptation. Ph.D. dissertation, University of North Carolina, Chapel Hill.
        Ann Arbor, MI: University Microfilms.
1977   Alternatives in Modernization: Styles and Strategies in the Acculturative
        Behavior of Baffin Island Inuit. Ethnography Series, ND 5-001. New
        Haven, CT: HRAFlex Books.
1980   The Politics of Inuit Alliance Movements in the Canadian Arctic. In
        Political Organization of Native North Americans. E. L. Schusky, ed.
        Pp. 243–282. Washington, DC: University Press of America.
1996   Modernization, Social Change, and Health in the Eastern Arctic. In Human
        Ecology and Health. Maj-Lis Foller and Lars O. Hansson, eds. Pp. 73–94.
        Goteborg, Sweden: Goteborg University, Section of Human Ecology.

Nuttall, Mark
1992   Arctic Homeland. Toronto: University of Toronto Press.

Pitseolak, Peter, and Dorothy Eber
1975   People from Our Side: An Eskimo Life Story in Words and Photographs.
        Bloomington: University of Indiana Press.

Riewe, R.
1991   Inuit Land Use Studies and the Native Claims Process. *In* Aboriginal
        Resources in Canada: Historical and Legal Aspects. K. Abel and
        J. Friesen, eds. Pp. 287–300. Winnipeg: University of Manitoba Press.

Stevenson, Marc G.
1997   Inuit, Whalers, and Cultural Persistence. New York: Oxford
        University Press.

Wenzel, G.
1991   Animal Rights, Human Rights: Ecology, Economy and Ideology in
        the Canadian Arctic. Toronto: University of Toronto Press.

6

Globalization, Dietary Change, and "Second Hair" Illness in Two Mesoamerican Cultures

George E. Luber

Global market forces have been affecting the health of Native Americans for over five centuries. Beginning with the conquest of what is now Mexico, the impact of these forces has ranged from igniting infectious disease epidemics of unprecedented proportions to the more subtle, but no less devastating, effects of poverty and malnutrition. The penetration of the global market economy continues to this day with the transformation of the few remaining small-scale, subsistence-based economies.

The subsistence farmers of southern Mexico's Sierra have always been at the periphery of the global economy. While astute marketeers, their contributions have historically centered on their supply of seasonal labor for coastal *fincas*, or plantations, and for large-scale government projects and their production of coffee and sugarcane for the export market. The move toward a cash-based local economy has hastened in recent years due to increased population pressure in the land-scarce highlands and the input of cash from relatives working in the United States. Not surprisingly, these shifts in the local economy, from subsistence agriculture to cash cropping and wage labor, have produced significant changes in the traditional lifestyle and transformed the epidemiological profile of these groups.

This chapter presents a case study of two ethnomedical syndromes, the Tzeltal Mayan *cha'lam tsots* and the Mixe *mäjts baajy*, which are regional variations of "second hair" illness found in several Mesoamerican cultures (Tenzel

1970; Lipp 1991; Heinrich 1994; Berlin and Berlin 1996; Luber 1999). I present data supporting the proposition that *cha'lam tsots* and *mäjts baajy* both represent ethnomedical diagnoses of protein-energy malnutrition, a potentially deadly form of malnutrition of which kwashiorkor and marasmus are extreme manifestations, and I hypothesize that the high prevalence of this illness in one of the study communities is the result of the residents' increased participation in the global economy and the subsequent commoditization of the local diet.

Glossed as "second hair" or "two hairs," *cha'lam tsots* is a Tzeltal Mayan ethnomedical syndrome identified by the presence of short, spiny hairs (one to two centimeters) growing close to the scalp, under the normal layer of hair (Berlin and Berlin 1996:279). It is a serious, potentially fatal condition that is believed to be caused by physical trauma to individuals, mostly children. Hair loss, diarrhea, fever, edema, loss of appetite, and general debility are its primary signs and symptoms (Figure 6.1).

An illness nearly identical to *cha'lam tsots* has been reported among the Mixe of Oaxaca, Mexico (Heinrich 1994; Lipp 1991). The Mixe *mäjts baajy*, or "two head hairs," primarily afflicts infants and is marked by diarrhea, anemia, a swollen body, puffed cheeks, and "numerous, fine shining hairs, or 'small

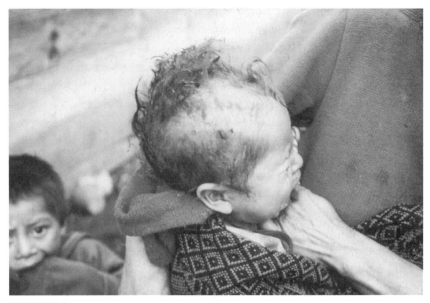

FIGURE 6.1
Tzeltal Mayan child with "second hair" illness.

spines,' growing on the head" (Lipp 1991:158). As with the Tzeltal Mayan *cha'lam tsots*, *mäjts baajy* is considered a serious and potentially fatal illness, primarily affecting children. Similar second hair illnesses have also been identified among the Cakchiquel Maya of Guatemala (Tenzel 1970) and the Jacaltecos and Motozintelcos of the Guatemala–Mexico border region (Mellado Campos et al. 1994).

In the following discussion, I present the ethnomedical perspective of second hair illness and develop its biomedical and epidemiological picture. I conclude with a discussion of the role of globalization in the emergence and prevalence of this illness in each study community. Although I had identified this illness in at least five southern Mexican indigenous groups, I decided to study this illness among the Tzeltal Maya of highland Chiapas and the Mixe of highland Oaxaca for two reasons. First, these two groups live in similar highland ecological settings, practice swidden (slash-and-burn) corn–bean–squash agriculture, and have similar levels of integration into the national economy. Second, the fact that this illness concept exists in linguistically distinct cultural groups points to the possibility that this concept crosses local boundaries and is shared by other groups in the Mesoamerican culture–area.

EXPLANATORY MODELS OF SECOND HAIR ILLNESS

To generate the ethnomedical description and explanation of each illness, I used the explanatory model methodology set forth by Arthur Kleinman (1980) and modified by Elois Ann Berlin and Brent Berlin (1996). Explanatory models (EMs) are sets of beliefs or understandings of specific illnesses that pertain to any or all of five domains: attributed etiology; diagnosis and typical onset; course, or evolution; prognosis, or predicted outcome; and healing strategy (Kleinman 1986:36). It has been widely used by medical anthropologists as a core technique in the cross-cultural comparison of illness concepts (Rubel and Hass 1990). As knowledge of these illnesses is widely shared throughout these regions, EM interviews were conducted with both specialist healers and lay persons with experience with this illness.

The following discussion focuses on describing the Tzeltal and Mixe EMs for second hair illness. As there is often variability between individual EMs, this analysis focuses on the principal shared components of the model, the most salient features, as elucidated by frequency of reports by the informants. These EMs are representations of the cognitive structures that persons invoke when thinking about, and acting on, an illness. They provide a useful framework for the cross-cultural understanding of an illness, and they illuminate points for comparison and for the exploration of biomedical equivalence. In general, the Tzeltal and Mixe EMs contain a good deal of information, derived primarily from the close and careful observation of this disease's pathological basis. Luber (2002) provides a detailed discussion of the correspondence between the EM and the biomedical model of second hair illness.

The Tzeltal Mayan *Cha'lam tsots*

The Tzeltal-speaking municipality of Tenejapa, located 25 kilometers east of San Cristóbal de las Casas, the regional center, comprises approximately 27 thousand inhabitants, located in 21 villages (*parajes*) and one municipal center (*cabecera*). The highland Maya economy is largely subsistence based, relying on corn, bean, and squash milpa agriculture, with seasonal plantation labor and coffee-growing becoming increasingly important (Cancian 1992). This region of the central highlands of Chiapas has been fertile ground for anthropological research and, as such, boasts a rich ethnographic and medical anthropological literature (for the former, Collier 1975; Cancian 1992; Vogt 1969, 1976; for the latter, Berlin and Jara 1993; Berlin and Berlin 1996; Menegoni 1996; Granich et al. 1999). As with much of rural Mexico, this region has seen dramatic socioeconomic and political changes over the last few decades (Cancian 1992).

With the help of a local collaborator, EM interviews were conducted with 7 traditional healers and 18 laypersons in the municipality of Tenejapa and neighboring Oxchuc. Knowledge of this illness is widespread, as most people, from farmers to taxi drivers, either have personal experience or know someone who has had this illness. Core elements of the EM for *cha'lam tsots* were strikingly homogenous. The only significant differences between specialist and lay models were in the domain of treatment options.

Attributed Etiology

The Tzeltal consider *cha'lam tsots* to be closely associated with diarrheal disease and ascribe to it a naturalistic etiology—that is, an illness that has a natural, not supernatural, causative agent. Infants and children can get *cha'lam tsots* through two principal means, each often resulting from the neglect of young and inexperienced mothers. First, *cha'lam tsots* can be caused by a blow to a child's head, such as when one falls down. This frequently happens to infants of 12 to 18 months who are just learning how to walk. The blow to the head explains the appearance of the short, spiny hairs and also the total loss of hair, which is a secondary characteristic of this illness.

Infants can also fall ill with *cha'lam tsots* when they are left out in the sun too long—for instance, by a mother who is working in the milpa or cornfield. According to this explanation, the intense sun and heat burn the head of the child, causing the new hair to grow differently and the old hair to fall out. After this incident or a series of such incidents, the child begins to weaken; the second type of hair appears; and the illness sets in. Exposure to heat and sunlight sufficient to cause *cha'lam tsots* can only occur during the dry months of the year, from March through early June, as there is too much cloud cover and rain the rest of the year. The following testimony, from two speakers, illustrates the nature of these traumas:

Sometimes it begins when they are little and they fall down and hit their head. Because of this, they say, the sickness *cha'lam tsots* comes.

Because [the child] doesn't want the heat when we harvest. And we have to carry the child to work . . . just so it arrives on the head.

Analysis of these ascribed etiologies highlight two important facts: this illness afflicts infants of the weaning age (12–18 months) and children during the dry months of the year, from March through early June.

Diagnosis and Typical Onset

The central sign for *cha'lam tsots* is the appearance of short, delicate, discolored, and spiky hairs on the head of the infant, under the normal layer of hair. One cannot be diagnosed with this illness without this sign. *Cha'lam tsots* translated means "two, or second, type of hair" (*cha* = two, *lam* = hairlike objects, *tsots* = hair). Although specialist diagnoses are important in Mayan ethnomedicine, nonspecialists are able to provide a positive diagnosis based on this criterion, though many often include diarrhea in the set of minimal criteria.

Although one cannot diagnose this disease absent the characteristic short hairs, several signs and symptoms can alert the caregiver that the infant may indeed have it. One of the earliest symptoms to emerge is the child's loss of appetite, as exemplified by the following statement: "They don't eat. They drop little pieces of their tortilla all over the ground." After the loss of appetite, the child begins to lose energy and appear irritable. This *debilidad* brings with it sleeplessness, bad dreams, and constant crying.

Preliminary signs of *cha'lam tsots* include mild diarrhea, edema, cough, fever, mild alopecia, and a reddish-colored rash on the head.

Course of the Illness

As the illness progresses, the preliminary signs and symptoms increase in severity. Specifically, the infant has constant diarrhea, continued weight loss, and increased edema. The edema typically begins in the face and progresses to the forearms and lower legs. Gastrointestinal and respiratory infections become more severe and a night fever is common. The most striking aspect of advanced cases of this illness is the severe alopecia, and most consider a case of *cha'lam tsots* to be advanced and life threatening when hair loss becomes significant. As the hair falls out, the scalp hairs are replaced by the fine, delicate, and light-colored "second hairs," the characteristic that gives this illness its name.

Prognosis

The most agreed-on element of the EM for *cha'lam tsots* was the set of re-sponses pertaining to the prognosis, or predicted outcome, of this illness. If patients do not seek treatment, they invariably die. One response during the interview included, "They die if they don't find the medicine of that plant. One dies, a person dies. If he finds the medicine, he doesn't die—only their head hurts them." Furthermore, two variant descriptions illustrate how a per-son might die from this illness: "One dies [because] one gets thin" and "One dies because it causes swelling."

Healing Strategy

Interviews revealed that only one treatment existed for *cha'lam tsots*. This treatment was generally well understood and agreed on by laypersons and specialists. As this illness has a naturalistic etiology, treatment is plant based, does not involve spiritual elements, and can be administered by a specialist healer or a layperson with specific knowledge of the treatment.

The healer first obtains the medicinal plant or plants to be used in the treatment. Several plants have been named as treatments for *cha'lam tsots*. The most commonly reported are *Chimaphila macolata, Verbena carolina,* and

Sambucus mexicana. The plant or plants are ground to a poultice, mixed with water, and applied cold to the scalp of the sufferer once a day for three consecutive days. The head of the patient is kept wrapped in a cloth for the duration of the treatment. Healers report high success rates with these treatments.

The Mixe *Mäjts Baajy*

The highland Mixe, or *Ayuuk jä'äy*, inhabit the high cloud forests of the Sierra Juarez in the southern Mexican state of Oaxaca. The Mixe are fiercely independent people with a strong shared sense of identity (Nahmad Sittón 1994). Like the Tzeltal Maya, the Mixe practice subsistence agriculture, focusing primarily on corn, beans, squash, and potatoes, and have been growing coffee and citrus for the export economy. As with the highlands of Chiapas, this region has undergone dramatic changes over the last 30 years in social and economic life, primarily due to the influences of out-migration to Mexico City and the United States.

With the help of a local collaborator and translator, EMs were elicited from 9 traditional healers and 14 laypersons in the highland municipalities of Totontepec, Tlahuitoltepec, and Mixistlan de la Reforma. Although knowledge of this illness was not widespread, those who had experience with it generally shared a high level of agreement.

One of the most significant findings involved the low prevalence of this illness. Without exception, all of our interviewees were surprised that we were asking about their knowledge of such a rare illness, and the interviewees often paused to remember certain facts about it. Our informants recollected that this disease used to be a common illness in the Mixe area but that it had disappeared in the mid-1970s (for reasons discussed later). This is in sharp contrast to the Tzeltal sample, where almost everyone had at least some idea of what this illness is. Additionally, a good deal of time was spent visiting outlying *agencias*, or hamlets, trying to track down a case of *mäjts baajy*, with no success. Data from the lowland Mixe populations corroborate this finding (Michael Heinrich, personal communication, January 2002).

Attributed Etiology

The Mixe explanation of causation for *mäjts baajy* is markedly different than the Tzeltal explanation of *cha'lam tsots*. Two basic models at work here, depending largely on the age of the healer. Older specialists (over 50 years old) typically reported that *mäjts baajy* is a particular manifestation of fright illness (*susto*), which is a common folk illness in Mexico.

One becomes *asustado* after a frightening experience that startles the blood and heart or dislodges one of the souls that a person accrues during one's lifetime. The loss of a soul is akin to losing part of one's life force, and children are especially vulnerable. *Susto* is brought about through frightening dreams that cause sudden movements in the child's arms and legs. "They jump in their bed, and we know that it is *susto.*" *Susto* weakens the body, rendering it susceptible to *mal aires* (bad airs), which can lead to a range of illnesses, especially *mäjts baajy.*

Most of the younger healers have been influenced by biomedical models of disease causation and therefore explain *mäjts baajy* as a nutritional disorder. "One gets it from a lack of eating" was commonly reported. Our informants reported that this malnutrition typically comes about because the mother did not care well for her child. The typical sufferer of *mäjts baajy* is from 12 to 18 months old, although children up to the age of six or seven are susceptible.

Diagnosis and Typical Onset

As with the Tzeltal *cha'lam tsots*, the diagnosis of *mäjts baajy* is contingent on the presence of short, delicate, spiny hairs appearing first on the back of the neck and then progressing to the whole head. Called *espinillas* ("little spines"), these delicate hairs exhibit a range of colors, from reddish to blonde, which signal the severity of the illness. The lighter the color, the more severe the illness. Preliminary signs of *mäjts baajy* include weight loss, facial edema, diarrhea, fever, sleeplessness, and irritability. Sunken eyes and a jaundiced, anemic complexion were also commonly reported.

Course of the Illness

As with the Tzeltal EM, the progression of the illness is first signaled by the spread of the *espinillas* down the patient's spine to the lower back and over the head. As the hairs spread, the child's condition deteriorates; diarrhea becomes more severe; weight loss quickens; the edema spreads from the face to the extremities; hair loss increases; and fevers become more common. Many reported a distended belly as the illness becomes severe. Psychologically, the child becomes increasingly lethargic and irritable as appetite all but disappears. The terminal phase of this illness is marked by total apathy to one's surroundings.

Prognosis

Again, agreement on prognosis was high. All informants reported that without treatment a sufferer of *mäjts baajy* would die. Most reported that a child would die because one did not want, or was unable, to eat: "The child will become thin because they don't want to eat. And then they will die. If they don't receive the treatment, they will die."

Healing Strategy

One of the most striking differences between the Tzeltal and Mixe EMs is in the treatment strategy. Both cultures agree that the second hairs are the pathological agent in these illnesses, and both treatments target the new type of hair growing on the scalp. While the Tzeltal treat the hairs and scalp by applying medicinal plants, the Mixe attempt to cure this illness by removing the offending hairs by shaving them off.

Once a child has been diagnosed with *mäjts baajy*, one is brought to a specialist healer. The healer begins the treatment by confirming the presence of the hairs. The hairs are revealed by spitting into the palm of a hand and rubbing the neck and spine until the spiny hairs stick up in the same direction. Once their presence is confirmed, the neck and back are rubbed with lard, and the affected area is shaved and cleaned with alcohol. Following this treatment, the child is given dried fish, beef, pork, or any "greasy" meat that is available. This dietary prescription is another important difference between the Mixe and Tzeltal models for these illnesses and is a result of the Mixe acknowledging a nutritional basis for the illness.

THE BIOMEDICAL AND EPIDEMIOLOGICAL PERSPECTIVE OF SECOND HAIR ILLNESS

As mentioned, use of the EM methodology was intended to develop avenues for exploration in the search for etiologic agents, risk factors, and biomedical equivalence. Once this ethnographic component was completed, I began collaboration with Dr. David Keifer, who is based in the United States, to explore the clinical and epidemiological picture of this illness. Based on the EMs, our hypothesis was that these illnesses most resemble several forms of protein-energy malnutrition, and our research methodology reflect this hypothesis.

Protein-Energy Malnutrition

The term *protein-energy malnutrition* (PEM) was suggested by Derrick B. Jelliffe (1966) to describe the spectrum of macronutrient deficiency syndromes, including marasmus, kwashiorkor, and nutritional stunting or dwarfism, which are caused by an inadequate dietary intake of protein and calories. The most common victims of PEM worldwide are children. There are two basic types of PEM: marasmus and kwashiorkor. Marasmus is caused by a decreased energy intake relative to energy expenditure and typically develops over a long period (Hensrud 1999). As marasmus advances, fuel stores are depleted, and the individual develops the characteristic wasted appearance. The hallmarks of marasmus are reduced body fat and lean tissue stores. Essentially, marasmus is starvation.

The term *kwashiorkor* comes from the Ga language of West Africa and can be translated as "disease of the displaced child" because it was commonly seen after weaning (Klein 2000). It refers to an inadequate protein intake with a fair or normal caloric intake (Balint 1998). Compared to marasmus, kwashiorkor can develop quite quickly, in a matter of weeks, without adequate nutritional support and in the presence of infection. The presence of peripheral edema is the hallmark sign distinguishing children with kwashiorkor from those with marasmus. Other signs suggestive of kwashiorkor in children include poor growth, decreased subcutaneous fat, muscle wasting, abdominal distension, dyspigmentation of the hair, easy pluckability and breakability of the hair, thin–sparse hair, diffuse depigmentation of the skin, and facial edema (Jelliffe 1966; Balint 1998). Children with kwashiorkor are typically lethargic and apathetic when left alone and irritable when picked up or held.

Recent research indicates that kwashiorkor might not always be caused by a relative deficiency in protein intake, as had been thought, because protein and energy intake can often be similar in children with kwashiorkor and marasmus (Klein 2000). Kwashiorkor, in these cases, is caused by the physiologic stress of an infection that induces a deleterious metabolic cascade in an already-malnourished child. Therefore, kwashiorkor is often an acute illness when compared with chronic undernutrition. This fact underscores the important two-way relationship between infectious disease and malnutrition.

The diagnosis of PEM is obvious in its most severe forms; in mild to moderate forms, the usual approach is to compare the body weight for a given height with that of standardized growth charts (Cotran 1999). From an anthropometric standpoint, the diagnosis of PEM is based on low body weight for height and on measures of subcutaneous fat (from triceps skinfold) and muscle (from midarm circumference) below the standard described in Frisancho (1990). The type of PEM can be identified based on the presence or absence of edema as well as on the child's weight in comparison with that expected for one's age (see Balint 1998).

Sampling and Methodology

As *mäjts baajy* has reportedly been eradicated from the Mixe region, the clinical picture of second hair illness comes from cases identified among the Tzeltal Maya. To locate cases of *cha'lam tsots*, we selected four villages in the Tzeltal Mayan municipality of Tenejapa to sample for the illness. With the aid of local collaborators, we searched each village for cases of *cha'lam tsots*. Door-to-door surveys, village announcements, and word of mouth were our primary means of locating cases. Over a one-month period (November 2000), we identified 19 cases (7 males, 12 females). With an estimated population of 35 hundred persons for these four villages, the prevalence of *cha'lam tsots* is 5.4 cases per 1,000 people.

For each case encountered, we conducted several different interviews and examinations, including a medical history, complete physical examination, collection of blood and stool samples (to test for parasites and serum albumin), and anthropometric measurements (weight, height, midarm circumference, triceps skinfold, head circumference, and chest circumference).

Findings

Through analysis of the clinical examinations, medical histories, and anthropometric data, our findings support our hypothesis that *cha'lam tsots* can be defined biomedically as PEM. All 19 cases were diagnosed with PEM. Although, as the EMs of *cha'lam tsots* and *mäjts baajy* clearly point to kwashiorkor as the type of PEM we expected to encounter, we were quite surprised that we were only able to diagnose two cases of kwashiorkor among our samples.

Although edema was a key component of the EM for *cha'lam tsots*, only three of our cases presented edema. We did commonly observe, however, other clinical signs indicative of PEM. All patients presented, to varying degrees, the short, discolored, spiny hairs described in the EMs. Several patients had advanced alopecia, and one infant, a 16-month-old girl, had lost almost all of her hair in the span of several weeks.

The anthropometric data provided the clearest picture of the nutritional status of these individuals. All but one of the 19 cases presented weights for age below the 50th percentile; 13 of the 19 were at or below the 5th percentile for weight for height. Similar observations were made for height: 14 of the 19 cases were at or below the 5th percentile, and no case was above the 50th percentile. Using the standard of reference described in Frisancho (1990), we observed below-average to severe muscle wasting in 8 of the 19 cases; 10 of the 19 cases showed below-average measures of upper-arm fat.

Laboratory analysis of the ten cases that consented to blood draws and stool samples indicated no cases with serum albumin below normal levels (i.e., less than 3.5 grams per 100 milliliters).

One difficulty in identifying ethnomedically "typical" cases of *cha'lam tsots* is presented by the chronic nature of this illness. Many case histories reported that the illness would be severe for short periods, then broken by longer periods of remission, where the child's symptoms would improve. This was clearly the case with most of our sample, as caregivers reported in the medical histories that their children's conditions improved and that symptoms were less severe. An ethnomedical diagnosis of *cha'lam tsots* would therefore catch patients during both the acute and the remission phase of their illness.

As gastrointestinal infection, as evidenced by severe diarrhea, is a core criterion for a diagnosis of *cha'lam tsots*, one probable reason for the chronic nature of this illness might lie in the physiologic stress that infection induces in the already-malnourished child (Klein 2000). Fourteen of 19 cases reported moderate to severe diarrhea, while 10 cases reported upper respiratory tract infections. Few patients showed parasitic infection: only 3 had *Ascaris lumbricoides* infection. The severity of these symptoms covaries with the severity of one's overall condition. In these cases, there is a strong connection between overall nutritional status and the severity of one's gastrointestinal or respiratory infections.

Although we were unable to locate anyone with an ethnomedical diagnosis of *mäjts baajy* in Oaxaca, we were fortunate to interview a Mixe traditional healer with some perspective on this issue. Eloisa Flores, who had received extensive training in nursing at a Salesian monastery in Tlahuitoltepec, reported treating scores of *mäjts baajy* cases during her years at the Salesian clinic. During one of our interviews, I asked her to describe the appearance of a typical patient with *mäjts baajy*. She described the classic signs of kwashiorkor. Not satisfied, I kept probing, trying to narrow down the signs and symptoms until she, tiring of my questions, interrupted, "You know, it's malnutrition; the child has kwashiorkor." When I asked her how she was sure that it was kwashiorkor and not marasmus, she proceeded to describe accurately the differential diagnosis of the two conditions. Other than the EMs, this is the only available evidence on the potential biomedical equivalence of *mäjts baajy*. Considering the credibility of the source, we are confident that *cha'lam tsots* and *mäjts baajy* are, from a biomedical standpoint, the same disease.

THE POLITICAL–ECOLOGICAL ROOTS OF SECOND HAIR ILLNESS

One of the most interesting and unexpected findings to emerge from this research was the striking difference in prevalence of *cha'lam tsots* in Chiapas and *mäjts baajy* in Oaxaca. Although I expected *mäjts baajy* to be a rare illness, I did not expect to be unable to locate a single case. Conversely, I was surprised that so many Tenejapanecos were suffering from *cha'lam tsots*. In an attempt to uncover possible reasons for the discrepancy in prevalence, I interviewed regional epidemiologists from the Mexican National Institute for Social Security (Instituto Mexicano del Seguro Social; IMSS); Salesian nuns who ran a clinic in nearby Tlahuitoltepec that was responsible for early improvements in Mixe health; and village *promotores de salud*, who are local health care promoters and auxiliaries to larger health care projects. Perhaps most important, I interviewed local elders and municipal leaders to gain insights into some of the important changes that have taken place in the past 30 years.

Initially, I thought that it was fairly obvious why the Tzeltal were suffering from chronic and severe PEM (discussed in the following section), but I was somewhat perplexed as to how the Mixe, an indigenous group with similar levels of poverty, had rid itself of this illness. More important, the difference in prevalence offered a prime opportunity to explore some of the political–economic and ecological causes of PEM in Mexico's indigenous communities.

The following discussion offers an explanation of the reasons why *cha'lam tsots* is common in the Tzeltal region and how *mäjts baajy* had been eradicated by the Mixe. In the most general sense, two basic reasons account for the difference in prevalence I observed: the failure of public health projects in Chiapas; and the influence of dietary delocalization, fueled by global market forces, on the Maya diet in Chiapas.

Why Public Health Projects Work in Oaxaca and Fail in Chiapas

Efforts to improve health, sanitation, and nutritional status have met with differing degrees of success in Tzeltal and Mixe communities. Despite similar levels of poverty, marginalization, and degree of integration into the Mexican national economy, the Mixe nutritional profile is quite different from that of the Tzeltal. Much of the differences between the Mixe and Tzeltal health profiles can be attributed to the local success of a few health promotion and development projects.

Mixe have received, and have been receptive to, a number of development and public health projects that have improved basic sanitation and hygiene levels in the Mixe communities, even in the more remote and outlying *agencias*. Mixe historian Juan Areli Bernal Alcántara (1991) provides an overview of the earliest development programs, headed by the Salesian missionaries of the Roman Catholic Church. In 1972 they established a mission hospital in the western Mixe municipal center of Tlahuitoltepec and directed their efforts at improving hygiene, providing safe drinking water, and educating mothers on proper infant nutrition. Many older informants reported that the efforts of the Salesians had a significant impact on the living conditions, hygiene, and nutritional status of the highland Mixe.

The building of roads into the Mixe highlands during the early 1970s brought a dramatic change in Mixe health. During this time, the National Indigenous Institute (Instituto Nacional de Indigenista; INI) established the first medical posts in many villages. These clinics, staffed by full-time nurses, provided basic health care and carried out vaccination programs.

In the early 1980s, the IMSS followed up on the early success of the INI and Salesian programs with the COPLAMAR program (Coordinación General del Plan Nacional de Zonas Deprimidas y Grupos Marginados, "General Coordination for the National Plan for Deprived Zones and Marginalized Groups"), which built a network of rural clinics, each staffed by a full-time physician and nurse. These clinics provide the bulk of Western medical care in this region and carry out nutritional screenings, prenatal exams, disease prevention workshops, and even family planning clinics (Lipp 1991). Each clinic is equipped with a pharmacy and an operating table to handle emergencies.

Building on the early success of the efforts of INI and the Salesians, investigators at CECIPROC (Centro de Capacitación Integral para Promotores Comunitarios, "Center for Integrated Training for Community Workers"), a project of the Instituto Nacional de la Nutrición (National Institute of Nutrition) "Salvador Zubirán," developed a program in the mid-1970s to promote health, sanitation, and child nutrition in the highland Mixe area (Diez Urdanivia and Ysunza Ogazón 1996). These researchers discovered high rates of protein malnutrition in the highland Mixe communities and focused their efforts on both the assessment of nutritional status and the integration of protein-rich foods into the Mixe diet.

Due to the difficulty of obtaining high-quality proteins in this remote location, CECIPROC focused initially on improving access to dairy and poultry products by supplying chicken and turkey chicks to Mixe households, often without cost. Their early efforts met with little success, and in an evaluation of the project, they realized that the Mixe did not know how to cook with cheese, eggs, and milk, which were to provide the bulk of the animal proteins (Ysunza Ogazón et al. 1993; Armando Sánchez Reyes, personal communication, May 2001). They developed community education seminars where they would teach women how to use these "new" foods and incorporate them into their diet (Diez Urdanivia and Ysunza Ogazón 1996). Despite these initial difficulties, CECIPROC has been quite successful at integrating milk, cheese, eggs, and poultry into the traditional diet (Yzunza Ogazón et al. 1991; Sesia 1993).

My own observations, based on my living in these communities for several months, confirm these results. Rarely would a meal be served that did not have cheese, beans, poultry, or even beef. The local favorite *tlayuda*, a large tortilla filled with salsa, *epasote*, and other herbs, is now served with melted cheese inside, making it a sort of quesadilla. Fried cheese is commonly served for breakfast, especially to children. Eggs are common in all meals and are no longer imported to the communities but are raised locally. I have spoken with several elderly Mixe women about these dietary changes, and they all confirm that since the arrival of the roads in the early 1970s, the Mixe diet has been transformed and this transformation has improved their nutritional status.

The situation in Chiapas is quite different. No development projects have had the type of impact on health and nutrition that is evident in the Mixe region. Ethnoepidemiological surveys of the highland Maya undertaken by Berlin and Berlin (1996) reveal that gastrointestinal disease and respiratory infections form the bulk of illness episodes. This is not surprising given the high levels of parasitism and poor sanitary conditions characteristic of these communities. Malnutrition is also a serious problem, especially in the Tzeltal and Tzotzil communities. Indeed, nutritional surveys in the highlands of Chiapas have demonstrated that the Mayan diet suffers from several deficiencies, especially animal protein (Perez Hidalgo 1975; Berlin 1999).

Health services have developed slowly in highland Chiapas (Menegoni 1996). The first state-supported health care programs were instituted in the early 1950s by the INI, which opened 13 health posts and clinics in Tzeltal and Tzotzil communities and trained local health promoters in basic Western medical techniques. The IMSS–COPLAMAR and the Coordinated Health Services established 56 health clinics in highland Chiapas beginning in the 1970s, but, as in the Mixe region, the government-run programs are largely directed at curative rather than preventative services (Berlin and Berlin 1996). Some of these programs—IMSS–COPLAMAR and, more recently, PRO-GRESSA (Programa de Educación, Salud, y Alimentación, "Program for Education, Health, and Food")—have been working to reduce the high levels of gastrointestinal disease by improving personal hygiene, promoting the use of latrines, and providing access to safe drinking water. Still, 38% of our study population did not use latrines.

The Tzeltal diet suffers from a lack of quality proteins. In a recent inventory of most commonly consumed foods, Berlin and colleagues (1998) discovered that eggs and milk were the most commonly consumed high-protein foods, at only seventh and eighth in frequency of overall use, behind coffee, sugar, and even soft drinks. Efforts to integrate cheese into the Tzeltal diet have met with little success (Berlin and Jara n.d.). My personal experience living in Tzeltal communities supports these findings. Milk, cheese, and eggs are not part of the daily diet, and poultry and meat products are eaten only on ceremonial or important occasions.

A new project aimed at improving the nutritional status of poor Mexicans has been initiated by IMSS and Coordinated Health Services under the name PROGRESSA. One of its principal weapons against malnutrition is a powdered supplement called *polvo de papilla* (powdered papilla). This powdered supplement was designed to provide extra protein, vitamins, and minerals to weaning-age children in an easy-to-use powdered formula. Evaluations of this program have revealed, however, that it has had little to no impact on the health and nutritional status of the program participants (Rivera et al. 2001:39).

One major reason why the program has not worked in Tenejapa is the lack of willingness of the program participants to utilize the nutritional supplements. Several Tzeltal in Tenejapa reported that, instead of feeding infants, the papilla was commonly used to fatten livestock prior to sale. The program designers hypothesized that the papilla was not used because the participants did not know how to use it. I spoke with several people about this practice, and they reported that the fattened livestock could provide extra money to the household economy and thus help in feeding the whole family. Investigators at CECIPROC reported that feeding the papilla to livestock was a problem in Totontepec before the implementation of educational programs aimed at teaching mothers on the proper use and benefits of the papilla supplement.

Topographical factors also play a role in the differences in the utilization and success of health promotion projects between the Mixe and Tzeltal. As discussed, the extreme vertical topography and lack of level ground necessitate that the highland Mixe live in concentrated settlements. In contrast, the gentler Tzeltal topography, and generally better soil conditions, favors a more dispersed settlement pattern. While the *parajes* are sociopolitically organized around a central plaza, typically containing the primary school, house compounds are widely dispersed, presumably to allow easier access to milpa plots. The compact settlement pattern of the Mixe has significance for the efficacy of public health interventions, as the majority of the populace is easy to reach and assemble. Clean running water, sewage, and drainage systems are also cheaper and easier to provide and maintain in concentrated populations, as is medical care.

In Totontepec, the majority of the population in the municipal center utilize the IMSS clinic because it is so easily accessible. Each *agencia* is visited weekly by the clinic doctor, who is able to see more patients because there is no need to travel long distances to visit individual homesteads. In contrast, many in Tenejapa have to walk several miles to see a doctor, a task made ever more difficult when one is sick.

Globalization, Dietary Delocalization, and the "Junk Food" Diet

Nothing weakens an immune system and overall health as efficiently as malnutrition, especially if families are, for economic reasons, substituting cheap fat and starch for more expensive proteins and fresh vegetables.

—L. Garrett, *Betrayal of Trust: The Collapse of Global Public Health*

The highlands of Chiapas have undergone a profound economic transition since the late 1970s. The increased importance of migration, seasonal wage labor, and the production of commodity crops such as coffee and sugarcane has had a significant impact on Tzeltal economic and social life and has altered traditional patterns of consumption (Cancian 1992). Among the facets of life most affected by this shift toward integration into the global, monetized economy has been the traditional diet.

The increased importance of export cash crops such as coffee and sugarcane leads to increased participation in the market economy. As peasant farmers allocate more of their production to cash crops, their use of costly inputs such as fertilizers and pesticides increases at a greater rate than their productive outputs. In essence, they exchange a stable low-cost, low-productivity agricultural system for a high-cost, high-productivity system in which their own returns continuously decrease as a result of unequal exchange in the market (Dewey 1979). The ever-decreasing global price of coffee is evidence of these diminishing returns. Several farmers interviewed in Tenejapa have begun to leave the beans on the plant due to the decreased price of raw coffee and the high cost of labor.

The dietary effects of the commoditization of the agricultural system are numerous. From an ecological standpoint, cash cropping reduces ecosystem diversity, and the reduction of fallow periods that accompanies cash cropping is detrimental to the soil quality. The loss of diversity directly affects the availability of wild uncultivated foods, which are an important nutritional resource for subsistence agriculturalists, especially at the end of the agricultural cycle and during famine. Most important, the shift away from subsistence production to cash cropping leads to the increased dependence on purchased foods (Dewey 1980; Pelto and Pelto 1983). This leads to poorer dietary quality and, ultimately, lowered nutritional status.

Dietary delocalization refers to the process in which there is an increase in the importance of foods from outside the region in the local diet (Pelto and Pelto 1983). While food commoditization and dietary delocalization are associated with improved levels of nutrition in industrialized nations (by increasing diversity), it has negative effects in developing countries (by reducing access; Dewey 1980; Pelto and Pelto 1983; Daltabuit and Leatherman 1998).

Dietary delocalization and the penetration of "junk foods" into the Mayan diet have also been well documented. M. Daltabuit and T. L. Leatherman (1998) discuss the influence of the tourism-led economy on the diet of Yucatec Maya. As integration into the monetized economy increases, milpa production decreases and diets shift away from a base in local produce to one in commercialized foods. Although these processed foods have increased caloric intake, they have not improved nutritional status. For school-age children, the authors report average weekly intakes of 7.4 soft drinks, 10.2 snack foods (cookies and chips), and 11.2 candies (326). In general, when compared with baseline data, the authors' posttourism data indicate increases in both child undernutrition and adult overnutrition, with consequent increases in obesity and diabetes.

The Tzeltal have embraced carbonated beverages as a substitute for *pox*, or cane liquor, which is an important element of all social and ritual activities. Although this has had a positive impact on levels of alcoholism, it has had a negative impact on infant nutrition, as most mothers see no danger in allowing their children to drink large quantities of this relatively cheap drink. It is not uncommon for weaning-age infants to drink up to a half-liter of soda per day.

Berlin and colleagues (1998) report that approximately half of the most frequently consumed foods in the highlands are nonsubsistence products. Soft drinks are the sixth-most consumed product in the Tzeltal diet, ahead of milk, eggs, and bread. My experience in Tenejapa supports the findings of Berlin and Berlin. Whereas foraging for wild foods used to be a common childhood activity, especially for boys, parents reported that children are now interested only in the occasional *peso*, which they can spend on soda and candies at the neighborhood *tienda*.

In a situation of limited economic resources, this pattern of consumption reinforces the diminishment of nutritional status, as consumption shifts from locally produced foods to national and global market products. Fueling these patterns of consumption are market forces and advertising that promotes the consumption of processed foods as markers of status. The increased preference of nonlocal foods, especially that for sodas and sweets, reflects a general shift in food choice from low-status, high-quality foods to high-status, low-quality foods.

As evidenced by the high rates of *cha'lam tsots*, severe malnutrition is a serious problem in the Tzeltal region. Although severe malnutrition is undoubtedly exacerbated by the high rates of gastrointestinal infections in young children, those forces that have altered the Tzeltal diet have been prime contributors as well.

CONCLUSION

The preceding discussion of second hair illness in southern Mexico has highlighted the important role that globalization has had in transforming the health and nutritional status of marginal groups. These changing epidemiologic profiles offer new challenges to public health efforts. While the focus in these regions has been, and continues to be, on improving access to sanitary water and sewage systems and on controlling infectious disease such as HIV and malaria, attention must also be given to changing nutritional status.

As this example illustrates, caloric intake can increase with subsequent malnutrition. Proper food choices must be promoted as part of the basic public health effort. Mothers must be educated about proper weaning foods for their infants, and adults should be informed about the dangers of diabetes and obesity, two epidemics that are taking hold in these communities.

While this example illustrates some of the consequences of economic integration into market economies, it is important to note that these same forces can and should be used to promote healthy behaviors. As the influence of global market media, through radio, television, and print advertising campaigns, is not likely to diminish, these same media should be used as conduits of public health messages. In conclusion, attention to the impact of globalization on the health of indigenous populations can not only reveal some of the root causes of disease but can also provide avenues to explore in the effort to alleviate some of the health disparities that are all too apparent.

NOTE

I would like to acknowledge the support and contributions of several people who have made this research possible: Dr. David Keifer, who provided his time and energy for the clinical examinations and evaluations; José Guzman Gomez, my friend and field collaborator for the Tzeltal Maya component of this research; and Areli Bernal Alcántara, friend, inspiration, and collaborator for the Mixe component. Funding for this research was provided by a Wenner-Gren Individual Grant for Dissertation Fieldwork, a National Science Foundation Dissertation Improvement Grant, and a Jacob's Research Grant from the Whatcom Museum (Bellingham, Washington).

REFERENCES

Balint, Jane P.
1998   Physical Findings in Nutritional Deficiencies. Pediatric Clinics of
            North America. 45(1):1–12.

Berlin, E. A.
1999   Unpublished data. Department of Anthropology, University of
          Georgia, Athens.

Berlin, Elois Ann, and Brent Berlin
1996   Medical Ethnobiology of the Highland Maya of Chiapas, Mexico:
          The Gastrointestinal Diseases. Princeton, NJ: Princeton University Press.

Berlin, E. A., O. B. Berlin, and G. E. Luber
1998   Biodiversity of the Highland Maya Diet. Paper presented at the American
          Anthropological Association Annual Meetings, Philadelphia, December 3.

Berlin, E. A., and V. M. Jara
1993   Me' winik: Discovery of the Biomedical Equivalence for a Maya
          Ethnomedical Syndrome. Social Science and Medicine 37(5):671–678.

Berlin, E. A., and V. Jara
N.d.   Patrones Etnoepidemiologicos en Comunidades Indigenas de los Altos
          de Chiapas 1: Ninos de 1 a 5 anos. PROCOMITH, A.C.:1–9.

Bernal Alcántara, Juan Areli
1991   El Camino de Anukojm—Totontepec y Los Salesianos. Totontepec, Mexico:
          Sociedad Cultural Totontepecana.
2001   Totontepec, Villa de Morelos, Mixe, Oaxaca: 20 Preguntas Basicas Acerca
          del Municipio. Totontepec, Mexico: Instituto Comunitario Mixe Kong Oy.

Browner, C. H., B. R. Ortiz de Montellano, and A. J. Rubel
1988   A Methodology for Cross-Cultural Ethnomedical Research. Current
          Anthropology 29(5):681–702.

Cancian, Frank
1992   The Decline of Community in Zinacantan: Economy, Public Life, and
          Social Stratification, 1960–1987. Stanford: Stanford University Press.

Collier, George Allen
1975   Fields of the Tzotzil: The Ecological Bases of Tradition in Highland
          Chiapas. Austin: University of Texas Press.

Cotran, Ramzi
1999   Protein-Energy Malnutrition. In Robbins Pathologic Basis of Disease.
          Ramzi S. Cotran, Vinay Kumar, Tucker Collins, Stanley L. Robbins, eds.
          Pp. 436–439. Phildadelphia: W. B. Saunders.

Daltalbuit, M., and T. L. Leatherman
1998   The Biocultural Impact of Tourism on Mayan Communities. *In* Building a
         New Biocultural Synthesis. A. Goodman and T. L. Leatherman, eds.
         Pp. 317–338. Ann Arbor: University of Michigan Press.

Dewey, K.
1979   Agricultural Development, Diet and Nutrition. Ecology of Food and
         Nutrition 8:265–273.
1980   The Impact of Agricultural Development on Child Nutrition in Tabasco,
         Mexico. Medical Anthropology (4):22–43.

Diez Urdanivia, Silvia, and Alberto Ysunza Ogazón
1996   Nutrición Comunitaria: Cuaderno de Trabajo I. Oaxaca, Mexico:
         Instituto Nacional de la Nutrición "Salvador Zubirán."

Frisancho, A. Roberto
1990   Anthropometric Standards for the Assessment of Growth and Nutritional
         Status. Ann Arbor: University of Michigan Press.

Garrett, L.
2000   Betrayal of Trust: The Collapse of Global Public Health. New York:
         Hyperion.

Granich R, M. F. Cantwell, K. Long, Y. Maldonado, and J. Parsonnet
1999   Patterns of Health Seeking Behavior during Episodes of Childhood
         Diarrhea: A Study of Tzotzil-Speaking Mayans in the Highlands of
         Chiapas, Mexico. Social Science and Medicine 48(4):489–495.

Heinrich, Michael
1994   Herbal and Symbolic Medicines of the Lowland Mixe (Oaxaca, Mexico):
         Disease Concepts, Healer's Roles, and Plant Use. Anthropos 89:73–83.

Hensrud, Donald D.
1999   Nutritional Screening and Assessment. Medical Clinics of North America
         83(6):1–15.

Jelliffe, Derrick B.
1966   The Assessment of the Nutritional Status of the Community. Geneva:
         World Health Organization.

Klein, Samuel
2000   Protein-Energy Malnutrition. *In* Cecil Textbook of Medicine. 21st ed.
         Russell Cecil, J. Claude Bennett, and Lee Goldman, eds. Pp. 1148–1152.
         Philadelphia: W. B. Saunders.

Kleinman, Arthur
1980   Patients and Healers in the Context of Culture. Berkeley: University of
        California Press.
1986   Concepts and a Model for the Comparison of Medical Systems as Cultural
        Systems. *In* Concepts of Health, Illness and Disease: A Comparative
        Perspective. Caroline Curer and Meg Stacey, eds. Pp. 29–47. Oxford: Berg.

Lipp, Frank
1991   The Mixe of Oaxaca: Religion, Ritual and Healing. Austin: University of
        Texas Press.

Lohman, Timothy. G., Alex F. Roche, and Reynaldo Martorell, eds.
1986   Anthropometric Standardization Reference Manual. Champaign, IL:
        Human Kinetics.

Luber, George
1999   An Explanatory Model for the Maya Ethnomedical Syndrome:
        Cha'lam tsots. Journal of Ecological Anthropology 3:14–23.
2002   The Biocultural Epidemiology of "Second-Hair" Illness in Two
        Mesoamerican Communities. Ph.D. dissertation, Department of
        Anthropology, University of Georgia

Mellado Campos, Virginia, Armando Sánchez Reyes, and Paolo Femia
1994   La Medicine Tradicional de los Pueblos Indígenas de México. Tomo II.
        Mexico City: Instituto Nacional Indigenista.

Menegoni, A.
1996   Conceptions of Tuberculosis and Therapeutic Choices in Highland
        Chiapas, Mexico. Medical Anthropology Quarterly 10(3):381–401.

Nahmad Sittón, Salomón
1994   Visión Retrospectiva y Prospectiva del Pueblo Mixe. *In* Fuentes Etnológicas
        para el Estudio de los Pueblos *Ayuuk* (Mixes) del Estado de Oaxaca.
        Salamón Nahmad Sittón, ed. Pp. 11–49. Oaxaca City, Mexico: Instituto
        Oaxaqueño de las Culturas.

Pelto, G. H., and P. J. Pelto
1983   Diet and Delocalization: Dietary Changes since 1750. Journal of
        Interdisciplinary History 14:507–528.
1990   Field Methods in Medical Anthropology. *In* Medical Anthropology.
        T. M. Johnson and C. F. Sargeant, eds. Pp. 269–297. New York: Praeger.

Perez Hidalgo, Carlos
1975   La Situacion Nutricional de la Poblacion Rural de los Altos de Chiapas.
        America Indigena 35:71–99.

Rivera, J. A., C. G. Rodriguez, T. Shamah, J. L. Rosado, E. Casanueva, I. Maulen,
    G. Toussaint, and A. Garcia-Aranda
2001   Implementation, Monitoring, and Evaluation of the Nutrition Component
        of the Mexican Social Program (PROGRESA). Food and Nutrition Bulletin
        21(1):35–42.

Rubel, A. J., C. W. O'Nell, and R. C. Collado-Ardon
1984   Susto: A Folk Illness. Berkeley: University of California Press.

Rubel, Arthur J., and Michael R. Hass
1990   Ethnomedicine. In Medical Anthropology: Contemporary Theory
        and Method. T. M. Johnson and C. F. Sargent, eds. Pp. 113–130.
        New York: Praeger.

Sesia, Paola
1993   Aportes al Estudio de la Medicina Tradicional y la Herbolaria en Oaxaca.
        Cuadernos del Sur 23(5):105–120.

Tenzel, James H.
1970   Shamanism and Concepts of Disease in a Mayan Community. Psychiatry
        33:372–379.

Vogt, Evon Z.
1969   Zinacantan: A Maya Community in the Highlands of Chiapas. Cambridge,
        MA: Harvard University Press, Belknap Press.
1976   Tortillas for the Gods: A Symbolic Analysis of Zinacanteco Rituals.
        Cambridge, MA: Harvard University Press.

Ysunza Ogazón, A., S. Díez-Urdanivia, and L. López Nuñez
1993   Programa de Investigacion-Accion Comunitaria en Migracion y Nutricion.
        Salud Publica de Mexico 35(6):569–575.

Yzunza Ogazón, Alberto, Fabiola Rueda Arroziz, Sara Pérez-Gil Romo, Silvia Diez-
    Urdanivia, and Laurencio Lopez
1991   Condiciones Materiales para la Salud y Situación Nutricional en Tres
        Grupos Etnicos Oaxaquenos. Revista del Instituto Nacional de la Nutrición
        Salvador Zubrian 3(14):8–21.

POPULATION DYNAMICS

7

Canadian Cases of the Public Health Implications of Global Environmental and Economic Change

John Eyles and Nicole Consitt

In this chapter, we examine some of the health consequences of globalization for Canada in the context of risk perception and the "culture of fear." We outline the relationships between health and globalization, particularly through Canadian examples of severe acute respiratory syndrome (SARS), West Nile virus, bovine spongiform encephalopathy (BSE; "mad cow disease"), and multidrug-resistant tuberculosis. While these conditions exist in other parts of the world, they may have particular salience in a country that is open to world forces through its trade and travel connectivity and immigration and refugee policies. Finally, we examine options for government and individual responses to these health consequences.

Globalization is often viewed as the flow across national boundaries of goods and services, capital, people, technology, ideas, and culture (U.S. General Accounting Office 2001; Labonte 2003). Often considered an inevitable outcome of modernization, its effects have been viewed as being both positive and negative, sometimes a matter of opinion and sometimes one in fact. What is agreed on, however, is that the world is experiencing rapid change, fueled by increased interconnectivity permitted by exponential technological developments. The ease of population movement and mobility, both in short-term travel and migration, has resulted in increased cultural diffusion as well as the spread of disease (MacPherson and Gushulak 2001). The case studies presented here exemplify how quickly disease can spread in a globalized world and what role the media and technology play in this accelerated dispersal.

RISK, FEAR, AND PUBLIC HEALTH

Modern technology has almost removed space and time as barriers to communication between countries. The emergence of international systems and institutions such as Microsoft and AOL Time Warner has created not only technological consequences but cultural ones too, including the homogenization of culture and consumption. Indeed, one key element in public and governmental response to these forces is the globalization of risk. In fact, current society is seen as being characterized predominantly by risk and the distribution of ills rather than the distribution of positives. Although the relative risk of SARS was small, for example, it dominated news headlines for weeks. Media channels attach themselves to events such as the SARS outbreak because they possess the elements of the unknown, invisibility, and dread impact. Furthermore, through the interconnectedness of the world through trade, travel, and migration, risk appears to be democratized (Beck 1992), with all the world appearing at risk and with too little protection to such airborne viruses.

Yet there also remain localized events and circumstances that may become risks with a particular coincidence of forces. Such a perceived glut of risks makes their management particularly challenging for public health agencies charged with ensuring safety and protecting health. Poorly understood diseases such as SARS and BSE are continuously emerging and defy easy answers and solutions. As a result, the public, who generally have little comprehension of science, receive equivocal information. Because knowledge is linked to our perceptions of risk, our fears are heightened by the science community's apparent inability to come up with answers and forms of treatment. Power and control are important mediators of risk perception and the experience of being unsafe (Beck 1996; Giddens 1998): a person's fear tends to increase when potential risks are imposed by others, thus influencing one's feelings of hopelessness and a loss of control. News media can affect our risk perceptions by socially amplifying the risk of public health events. As the SARS event has illustrated, the danger of new diseases are merely a plane ride away. Also, localized public health triumphs, such as the North American control of tuberculosis (TB), can be difficult to maintain as global population movements make health problems borderless. Taken together, these forces may amplify a "culture of fear." Barry Glassner (1999) has argued that public panics are produced by the media and thus lead to a culture of fear, characterized by exaggerated concerns over minor threats to health, safety, and well-being. These in turn cost large amounts of money, which are diverted from more serious public health problems, such as tobacco-related deaths or auto accidents.

People are more aware of health dangers now through the various media channels available to them. Thus, globalization has the tendency to reinforce these ideas of uncertainty. Media channels, in the form of Internet, e-mail, and satellite technology, have enabled fear to spread among populations faster than nature's rate of viral transmission. Indeed, as David Baltimore (2003) has stated, public anxiety over epidemics is outpacing medical technology's ability to cope with emerging diseases. And this uncertainty, fueled by disagreements within scientific research and an ever-changing knowledge base, has weakened public trust in modern institutions of "expert" knowledge (Giddens 1991). This nonagreement is exemplified by the literature on BSE, wherein discordance still exists regarding if BSE is the causal link for cases of new variant Creutzfeldt–Jakob disease in humans. In 2001, the *British Medical Journal* published an article disputing the link between BSE and human disease. George A. Venters (2001), who authored the highly publicized and controversial article, maintains that the infected-beef theory has not been supported by evidence to establish the link. The vehicle of BSE infection, a prion, has not been proven to be infectious in humans, nor has it been proven whether BSE can survive cooking, digestion, or the human immune response. Venters states that the evidence being collected aims to confirm the hypothesis, whereas a need to test the hypothesis is required in order to justify the media hype and public panic.

Uncertainty and heightened perceptions of risk create a market for information. In fact, a study on the media's role in the Canadian SARS outbreak reveals that the media played a dramatic role in shaping the crisis and reinforced the world's impression that Toronto's SARS outbreak was out of control. The authors of the study report that during the outbreak, local and national newspapers contained on average 3 to 6 SARS-related articles per day, with a peak volume of 25 articles (Feldman et al. 2003). While the probability of exposure to a particular pathogen is generally low, this point is typically given less media coverage than the potential consequences of risk to health. The enormous amount of media attention afforded to the recent outbreak of SARS can be compared with more prevalent diseases that typically receive much less attention. The current worldwide death toll due to SARS stands at 774, whereas 250 thousand to 500 thousand people die every year around the world due to ordinary strains of influenza (Finlay 2003). In the United States, which has available vaccines and medical care, influenza kills 36 thousand people every year (Centers for Disease Control and Prevention 2003a). Despite the media frenzy over the 2003–2004 flu season, health officials and doctors report that there is still no way of determining if any season's influenza is more severe than those in the past or if it is just off to an early start. The role of the media is key in any democratic society, but it bears a specific responsibility for reporting context—an element not seen in many public health issues (e.g., SARS, Ebola, BSE), especially given the importance of media news as an information source (in particular, that of television).

GLOBALIZATION AND HEALTH

The impacts of globalization are often most visible when discussing global health and the spread of disease. The World Health Organization (WHO; 1999) reports that infectious diseases kill 15 hundred people every hour, the majority succumbing to just six groups of diseases: HIV/AIDS, malaria, measles, pneumonia, TB, and dysentery and other gastrointestinal illnesses. Globalization is often cited as a primary driving force behind disease emergence and reemergence (see Armelagos and Harper, this volume). These factors include global travel, globalization of the food supply, population growth and urbanization, ecological and climate changes, and the emergence of drug-resistant microbes (Eyles and Sharma 2001). Close interactions resulting from global integration allows infectious diseases to spread rapidly, exposing populations to diseases to which their immune systems may not be accustomed. These factors are well described by Stephen S. Morse (1995) and outlined in Table 7.1. In some cases, such as that with SARS and that with the West Nile virus (described in the case studies presented here), the forces of globalization can allow the rapid spread of new diseases.

A specific example of this acceleration is the globalization of the food industry, characterized by the industrialization of food systems, factory farming, and long-distance transport of livestock. This increasing trend has been cited as a principal cause of the growing threat of animal diseases such as BSE for human populations. To increase production and associated profits, farmers have turned to new feeding practices and animal-processing methods. As a result of intensive farming and the associated close contact between animals, pathogens that are easily spread through livestock populations (e.g., *E. coli* 0157:H7, *Cryptosporidium bacterium*) have become endemic within the farm-

Table 7.1. Factors in Infectious Disease Emergence

Factor	Examples of Specific Factors	Examples of Diseases
Technology and industry	Globalization of food supplies; changes in food processing and packaging; organ or tissue transplantation; drugs causing immunosuppression; widespread use of antibiotics	Hemolytic uremic syndrome (*E. coli* contamination of of hamburger meat), bovine spongiform encephalopathy; transfusion-associated hepatitis (hepatitis B, C), opportunistic infections in immunosuppressed patients, Creutzfeldt-Jakob disease from contaminated batches of human growth hormone (medical technology)
Microbial adaptation and change	Microbial evolution, response to selection in environment	Antibiotic-resistant bacteria, "antigenic drift" in influenza virus
Breakdown in public health measures	Curtailment or reduction in prevention programs; inadequate sanitation and vector control measures	Resurgence of tuberculosis in the United States; cholera in refugee camps in Africa; resurgence of diphtheria in the former Soviet Union
Ecological changes, including those due to economic development and land use	Agriculture; dams, changes in water ecosystems; deforestation, reforestation; flood, drought; famine; climate changes	Schistosomiasis (dams); Rift Valley fever (dams, irrigation); Argentine hemorrhagic fever (agriculture); Hantavirus pulmonary syndrome southwestern U.S., 1993 (weather anomalies)
Human demographics, behavior	Societal events: Population growth and migration (movement from rural areas to cities); war or civil conflict; urban decay; sexual behavior; intravenous drug use; use of high-density facilities	Introduction of HIV; spread of dengue; spread of HIV and other sexually transmitted diseases
International travel, commerce	Worldwide movement of goods and people; air travel	"Airport" malaria; dissemination of mosquito vectors; rat-borne Hantaviruses; introduction of cholera into South America; dissemination of 0139 V. *cholerae*

Source: Morse 1995.

ing industry (Phillips 2001). As a result, toxic manure pools pollute human water supplies and result in tragic consequences (e.g., Walkerton and Battleford area outbreaks in Ontario and Saskatchewan) when circumstances related to heavy rains and complacency in water treatment are entered into the equation.

Factory feeding practices have been revealed as the chief cause of BSE contagion in livestock (Miller 1999). To increase profits through a source of cheap feed, livestock farmers began feeding cows rendered animal parts as a protein supplement. It is believed that this feeding practice allowed a prion agent to enter the cattle feed, and subsequent human consumption of these cattle instigated the human disease known as Creutzfeldt–Jakob disease. The 1986 outbreak of BSE in the United Kingdom was linked to this change in feeding practice; however, it was not until 1988 that feed restrictions were instituted. The U.K. outbreak of BSE infected nearly 20 thousand cattle and resulted in the slaughter of 4.5 million cattle (WHO 2002). The U.K. epidemic began in 1986 and peaked in 1992–1993, with almost one thousand cases reported per week. A 2003 King's Fund report stated that the media reporting of the BSE was handled poorly, inciting unwarranted fear and panic in the public. Journalists trumped up the human threat posed by BSE, conveniently forgetting that top research scientists warned ministers that the threat of BSE being transferred to humans was less than the chance of being struck by lightning (Harrabin et al. 2003). Presently, the BSE media hype in the United Kingdom has fizzled as the anticipated flood of new variant Creutzfeldt–Jakob disease cases has not materialized.

We have touched on some Canadian cases in these introductory sections. In the sections that follow, we provide additional Canadian case studies in more detail. We begin with two recent examples of diseases previously unknown to North America—SARS (2003) and West Nile virus (1999)—and we proceed with studies of diseases more familiar to Canadian soil, namely, TB.

CASE STUDIES FROM CANADA

SARS

The recent North American SARS outbreak illustrates the deadly alliance of infectious pathogens and the developments of globalization, both moving across the globe with the speed and efficiency of today's air travel. The first case of imported SARS arrived in Toronto from Hong Kong on February 23, 2003.

The virus showed up in Canadian hospitals just a few weeks later, with the majority of the 438 cases (251 probable and 187 suspect) occurring in Toronto. According to Health Canada (2003b), all of the 44 deaths attributed to SARS occurred in Toronto. In comparison, the province of British Columbia reported 50 cases of SARS (4 probable and 46 suspect). The comparison is interesting in that Vancouver and Toronto hospitals received their first cases of SARS within a week of each other: Toronto on March 7 and Vancouver on March 13. But Vancouver did not suffer a second outbreak of cases as did Toronto, where the outbreak spanned four months; put heavy pressures on public health and health care systems; and exposed weaknesses in national, provincial, and regional infectious disease surveillance capacities. The Toronto outbreak took root and established itself in hospital settings, where health care workers and staff, unaware of the new disease, exposed themselves and subsequent patient contacts to the infectious agent without proper protective barriers. Health care workers formed 108 of the 251 reported SARS cases.

Rapid transport, a phenomenon intimately associated with globalization, played a critical role in the spread of SARS. From its origins in Guangdong Province in South China, where it is suspected that an animal disease crossed to humans through dietary and farming practices, a doctor carried the virus to a Hong Kong hotel. There, the virus gained international mobility, as visitors at the hotel returned home or continued their travels. On March 26, 2003, SARS was declared a provincial emergency in Ontario, but by late April 2003, rates of transmission were slowing and control measures appeared to be working. Despite this, the WHO issued a rare emergency travel advisory on April 23 (WHO 2003b). This was the first SARS-related travel advisory issued for a Western country, creating added stigmatization of the Toronto area. Interestingly, the Centers for Disease Control and Prevention did not include Toronto on their list of travel-restricted areas, despite its proximity to the United States. These conflicting views regarding the safety of Toronto fueled public confusion and fear (Finlay 2003). During the last week of April, the virus peaked in what is now regarded as the end of the primary wave of infection. On May 14, 2003, WHO removed Toronto from the list of areas subject to local transmission. However, by May 23 Toronto announced that a second, nontravel-related cluster of cases had emerged, and by the end of May Toronto was once again on the WHO's "hot list" of recent local transmission (Health Canada 2003c).

On July 2 Toronto was removed from the list again, as the chain of local transmission had been halted and the outbreak contained. Worldwide, SARS cases totaled 8,098 by 2003, with a subsequent death total of 774 (WHO 2003a). There is widespread concern that SARS will become a seasonal disease, returning each year in the winter. The forces of globalization allowed the spread of this new disease faster than modern medicine could defend against, especially with weakened surveillance and control systems. Indeed, the world snuck in while Canada wasn't looking.

West Nile Virus

Before 1999, West Nile virus had never been detected in the Western Hemisphere; however, it has a well-established history of causing disease in Africa, West Asia, and eastern Europe. The virus, first isolated in Uganda in 1937, causes arthropod-borne disease and mortality in birds, mammals, and humans. West Nile virus emerged in North America in New York City during the late summer and early fall of 1999. This initial outbreak comprised 61 confirmed human cases of encephalitis, including 7 fatalities (Ministry of Health and Long-Term Care 2003b). Since then, West Nile virus rapidly established itself as a persistent public health concern, spreading throughout most of North America. Evidence that West Nile virus had arrived in Canada came in the summer of 2001, when active surveillance of the avian population indicated that several dead crows in southern Ontario tested positive for the virus. No human cases occurred in Ontario until 2002, when 319 confirmed and 85 probable cases occurred, most of which were distributed along the shores of Lake Ontario (Ministry of Health and Long-Term Care 2003b).

Viral activity in birds and mosquitoes increased West Nile virus presence in Ontario and resulted in its emergence in Quebec, Nova Scotia, Manitoba, and Saskatchewan (Health Canada 2002c). According to a recent study (Elliott et al. 2003), the seroprevalence of people who both lived in Oakville, Ontario, during the summer of 2002 and were infected by the West Nile virus is estimated to be 3.05% (CI = 2.18–3.92). This community infection rate was higher than expected and higher than the relative experience in the Queens borough of New York City in 1999, which had a seroprevalence of 2.6% (CI = 1.2–4.1). As of December 1, 2003, there were 852 probable human cases during the 2003 surveillance program, with 463 confirmed human cases and a total of 10 deaths (World Health Organization 2003a).

The rapid expansion of this vector-borne disease from Queens in 1999 to its eventual outbreak across virtually the entire continent of North America by 2003 exemplifies the speed with which our environmental and social systems can disperse disease (Nosal and Pellizzari 2003). In Ontario, the coincidence of particular weather patterns, trees, water sources and catch basins, and birds and their close proximity to human populations have been cited in the spread.

TB

Our last case examines the potential impact of travel and Canada's immigration and refugee policies on disease occurrence. Canada's TB rates are among the lowest in the world, yet new factors related to the emergence of multidrug-resistant strains of TB (MDR–TB) have highlighted the need for a sustained fight against TB in not only Canada but throughout the world. In 2002, TB among foreign-born individuals accounted for 66% of all reported cases in Canada, most of whom were emigrating from Southeast Asia, the Far East, sub-Saharan Africa, and the Indian subcontinent—all considered endemic regions. This is a dramatic shift when compared to the percentage of 1970, when foreign-born individuals accounted for 18% of TB cases in Canada (Health Canada 2003d).

Today, ease of travel and a shift in the demographic profile of migrants mean that immigrants may repeatedly visit their countries of origin, a practice not common in early migratory populations (MacPherson et al. 2001). Immigration from countries with a high prevalence of TB to countries of low prevalence has resulted in a global increase in the likelihood of encountering drug-resistant strains of TB (Long et al. 2003). MDR–TB is any strain of the bacteria that is resistant to the two most effective anti-TB drugs, isoniazid and rifampicin. It generally develops through improper use of TB medication, which is common in many non-Western countries. In 2001, 10.1% of patients diagnosed with TB in Canada were found to be resistant to at least one of the frontline drugs used to treat TB; 1% of the cases were of MDR–TB (Health Canada 2002b).

The government of Canada recently set an immigration target of 1% of the population (about 30 thousand people) per year. In 2000, the year of the most current national statistics, there were more than 90 thousand immigration landings in Toronto (40% of all new immigrants arriving in Canada that year; Pellizzari 2003). MDR–TB has been diagnosed with increasing frequency in

Ontario, especially Toronto. In 2001, Toronto Public Health had 369 new cases of TB. This represented 22% of all Canadian cases in 2001. Some 87% of the Toronto cases occurred in persons who were born outside Canada. Toronto has a rate of drug-resistant TB that is twice the Canadian average and a rate of MDR–TB that is two to three times higher than the Canadian average. With 3% of TB cases being MDR–TB, Toronto is considered by the WHO to be a hot area. A review of the 54 cases of MDR–TB that occurred in Toronto between 1995 and 2002 indicates that all but one occurred in persons who were foreign born (Pellizzari 2003).

Under current immigration protocols, an immigrant is required to undergo a medical examination (including a chest X ray) in one's country of origin before being admitted into Canada. Yet there is no integrated system for tracking those on medical surveillance, resulting in only 20% of such immigrants reporting to a local health unit within the stipulated 30 days (Uppaluri et al. 2002). Furthermore, latent TB infection can only be identified by tuberculin skin testing, which is not part of Canada's immigration protocols (Menzies 2003).

WHAT CAN BE DONE?

The fact that our cases are so wide-ranging and occur in a context of uncertainty and in a culture of fear gives the appearance of intractability. The cases share many features that play an important role in amplifying public perceptions of risk. They are new; they are potentially fatal; they are exotic; and they are poorly understood or are unfamiliar to the scientific community. When discordance emerges in the scientific community over the epidemiological profile of a disease, the media reports or exploits this nonconsensus, creating public skepticism and distrust in science and government. Yet control and treatment mechanisms do exist. Outbreaks of infections or newly emerging diseases have occurred in the past and have often been controlled with infrastructural investment (e.g., the development of safe water- and food-supply systems) or, as the last resort, measures of isolation and quarantine.

Quarantine is an important public health tool used to prevent those who are or may be infected with a contagious disease from infecting family, friends, coworkers, and the general public. Modern definitions of quarantine practices often distinguish between large-scale and smaller, individual-focused actions. Quarantine, by definition, entails the compulsory physical separation of populations who have potentially been exposed to a contagious disease (Centers for Disease Control and Prevention 2003c). Recently, the term *quarantine* has been (mistakenly) used to encompass many other public health measures, including travel limitations, restrictions on public gatherings, and isolation of physically sick individuals to prevent further spread of disease (Barbera et al. 2001). Quarantine practices evolved from the methods of the past that, in hindsight, can be deemed unethical. For example, quarantine was imposed on ships traveling to New York City from Europe during the 1892 cholera epidemic. Passenger immigrants arriving on these incoming ships were subjected to a two-tier quarantine system. Impoverished immigrants were sequestered below deck without basic sanitary provisions, while the passengers of higher socioeconomic status were not. Cholera spread rapidly throughout the lower deck, resulting in over 58 deaths on a single ship alone.

Canada has recently felt the consequences of government restrictions on commerce and transportation as a result of infectious disease. The international trade of Canadian beef was dealt a profound blow when mad cow disease was discovered in Alberta. Countries that import beef from Canada instituted a ban on shipments. The subsequent impact on the beef industry has been devastating, with many Canadian beef producers being on the verge of economic collapse. Although the intensive investigation that followed concluded with a finding of only a single case of BSE, the trade bans have been slow to lift in the three-dozen countries that introduced bans on Canadian cattle and beef products. The United States, which constitutes over 70% of Canada's international beef trade, along with Mexico and Russia, has slowly been opening its borders to Canadian beef (Yourk and Dunfield 2003).

In the case of SARS, the emergency travel advisory issued by the WHO essentially paralyzed the economic and tourist industries of Toronto. Fear and stigmatization remain, and the impacts have been far-reaching, extending beyond the Toronto area. It is estimated that the SARS outbreak has cost the tourism industry over CAN$350 million and will continue to effect losses into the near future. The net cost to the Canadian economy has been estimated to be between CAN$1.5 billion and CAN$2.1 billion (Health Canada 2003c). As well, the Chinese community in Toronto was marginalized during the outbreak because of generalizations made by society relating to the origins of the disease and its transmission. As the homosexual population was singled out when AIDS emerged in the 1980s, an anti-Asian sentiment spread across North

America as SARS became a racialized disease. Despite the fact that two-thirds of the cases were transmitted in a hospital setting, Toronto's usually vibrant Chinatown became deserted during the SARS outbreak (BBC News 2003b).

The SARS outbreak in Toronto resulted in over 25 thousand people being placed in, or asked to voluntarily submit to, a 10-day in-home isolation period. Reports of people refusing to abide by quarantine recommendations were frequently in the news during the outbreak, bringing into question whether health officials implemented strong-enough measures. In response to the public's noncompliance, Ontario health minister Tony Clement was quoted by a BBC news report (2003a) as saying, "We can chain them to a bed if that's what it takes." The reality is that enforcement was not strong enough, as exemplified by the second outbreak of SARS cases. Quarantine enforcement must always be considered with respect to such rights as liberty and freedom of movement. China raised international alarm among human rights groups when the Chinese Supreme Court announced in mid-May 2003 that those who chose to break quarantine measures would be subject to the death penalty or life imprisonment (Davis and Kumar 2003). Ironically, these extreme measures of containment followed a public-endangering government cover-up. By comparison, health officials in Ontario threatened to issue fines of up to CAN$5,000 a day or court-ordered isolation for those who chose to break SARS quarantine. Public health safety thus raises significant ethical issues.

Quarantine is an outcome of either on-the-spot decisions or surveillance. In fact, Michel Foucault (1977) argues that surveillance was developed in the 17th century to combat plagues through protocols and management technologies. For him, public health is a refinement of quarantine and as such is an instrument of the state. In fact, surveillance systems are by their nature designed to detect systematically infectious diseases through the monitoring and reporting of cases and the establishment of records and interventions. *Surveillance* is defined as "the ongoing, systematic collection, analysis, and interpretation of health data essential to the planning, implementation, and evaluation of public health practice, closely integrated with the timely dissemination of these data to those who need to know" (Thacker and Berkelman 1992). Yet surveillance capabilities have been seriously reduced in Canada both federally and provincially. Indeed, Laurie Garrett's (2000) comment rings true for Canada: "with a new century seeing the resurgence of old diseases alongside the emergence of new threats to health, effective public health infrastructures are more vital than ever—yet have been allowed to fall into disrepair." This was noted by the recent commission on public health after the SARS outbreak, which commented that Health Canada's presence was nonexistent and that provincial capacity had been stripped bare by government expenditure reductions and changed priorities (Health Canada 2003c).

Given public health's reduced capacity, response to these globally and environmentally related diseases is often exhortation—for voluntary quarantine or behavior change. Yet public perceptions of risk, while heightened by media and public health campaigns, are often accompanied by apathy toward specific hazards. For example, despite a CAN$38 million universal flu vaccination program established by the Ontario government, making flu shots free of charge to the entire population, the government reported an uptake of less than half the population. Yet 5 hundred to 15 hundred Canadians die each year from the flu or its complications (Health Canada 2003a). The burden of treating influenza is estimated to cost Ontario CAN$500 million each year, whereas the cost-effectiveness of vaccinating everyone who is eligible is projected to be CAN$320 million, or CAN$40 per vaccination (Health Canada 2003a).

Although awareness of the program has permeated 90% of Ontario's residents, uptake has been disappointing among the eight million residents eligible for the vaccination. In 2001–2002, there were 4.9 million doses of vaccine distributed, with an estimated 4.4 million administered (Ministry of Health and Long-Term Care 2002). A survey conducted to examine the underlying factors associated with vaccine uptake revealed that 48.3% of those who did not take advantage of the program cited that they felt as though they did not need it, because they were "healthy" and not at risk; 18.5% did not trust the vaccine; and 17.1% reported they did not have time. Between September 2001 and March 2002, Health Canada received 18 hundred reports of adverse reactions to flu vaccines, reactions that ranged from red eyes, respiratory symptoms, and facial swelling (Health Canada 2002a). Critics argue that the program is costing Ontarians CAN$44 million, plus another CAN$3 million in advertising, with little to show for this investment.

Similar apathy has been found with respect to West Nile infection. A study conducted for the Ontario Ministry of Health and Long-Term Care showed that while most area residents were aware of the risk of West Nile virus infection, as well as public health information about how to reduce the risk, area residents did not appear to undertake preventive measures as often as they could (Elliott et al. 2003). Similarly, a Connecticut-area study revealed that public belief in the presence of West Nile virus in its vicinity was not an established predictor for the use of personal protective barriers (Centers for Disease Control and Prevention 2003b).

This is not to suggest that governments should stop such public health programs or advocacy. But we live in a litigious society, distrustful of governments and armed with expansive civil rights. Canadians are no longer automatically respectful of institutional authority but rather are interested in their personal autonomy (Chartier and Gabler 2001). In light of this context, quarantine measures would likely be resisted. Furthermore, as demonstrated by the SARS outbreak in Toronto, placing 25 thousand people under quarantine created a public panic without benefiting the containment of the disease (Health Canada 2003c). It is now known that SARS is not highly contagious within the community; rather, it is spread through droplets and not readily airborne. In retrospect, focusing on screening, data management, and surveillance would have served control measures more effectively than the widespread quarantines enacted.

The SARS event highlights a dark side of globalization; however, it also serves as a lesson for the world and individual nations to update and institute measures of infectious disease surveillance and outbreak management. For example, amendments to the Canadian Quarantine Act in response to the SARS outbreak revealed legislation severely out of date. Many of the public health laws as they relate to quarantine were created before the developments in antibiotics and epidemiological tools. There is a need to review those laws. Not only does the public health legislature need to be updated, but Canada's public health infrastructure is chronically underfunded, as reiterated by the SARS commission. A seamless public health system for protocols and management is needed to integrate public health information and data among all levels of government. This coordination will create a cohesive public health environment that is prepared to respond to localized or national public health emergencies.

Any changes, however, will occur in a world of risk and fear. Risk is a dominant discourse in modern society (Ewald 1990) and, as our examples show, greatly influences the public's response to environmental and global change. This scenario has resulted in the development of risk-management protocols for decision making, although they may not always be used (Sparks 2001) since political and economic forces often determine courses of action. Furthermore, Canada's risk-management abilities may have been weakened by the reduction of the size and importance of government. In fact, David Garland (2003) argues that the welfare state was a risk-management state (see also, Whiteford and Hill, this volume). We expand his argument to suggest that its relative demise has serious implications for risk management. Furthermore, there will be tension between human rights advocates and those recommending the need for public action. It is possible that the World Trade Center tragedy has helped rationalize the need for a massive surveillance infrastructure. It has pointed to the need for worldwide cooperation and disclosure. These are indeed what proponents of global public health also advocate. Public health in Canada and throughout the world must reinvent itself and its existing tools to protect health, which is challenged by global influences of travel, movement, and the environment.

REFERENCES

Alexander, Diane L.
1997   Epidemiology of AIDS/TB Coinfection in Ontario—1990 to 1995. Public Health and Epidemiology Report Ontario 8(4):94–98.

Baltimore, David
2003   SARS Panic an Over-Reaction, Say Virus Experts. ABC News, April 29.

Barbera, Joseph, Anthony Macintyre, Larry Gostin, Tom Inglesby, Tara O'Toole, Craig DeAtley, Kevin Tonat, and Marci Layton
2001   Large-Scale Quarantine Following Biological Terrorism in the United States: Scientific Examination, Logistic and Legal Limits, and Possible Consequences. Journal of the American Medical Association 286(21):2711–2717.

BBC News
2003a   "Bed Chains" for Canada SARS Violators. BBC News, June 1. Electronic document, http://news.bbc.co.uk/2/hi/americas/2953496.stm, accessed December 2, 2003.
2003b   Chinatown's Virus Fears. BBC News, April 7. Electronic document, from http://news.bbc.co.uk/1/hi/health/2924399.stm, accessed December 5, 2003.

Beck, Ulrich.
1992   Risk Society: Towards a New Modernity. London: Sage.
1996   Risk Society and the Provident State. In Risk, Environment and Modernity: Towards a New Ecology. London: Sage.

Brower, Jennifer, and Peter Chalk
2003   Factors Associated with the Increased Incidence and Spread of Infectious
          Diseases. *In* The Global Threat of New and Reemerging Infectious
          Diseases: Reconciling U.S. National Security and Public Health Policy.
          Santa Monica, CA: RAND.

Centers for Disease Control and Prevention
2003a   Key Facts about the Flu. Electronic document, from
          www.cdc.gov/flu/pdf/keyfacts.pdf, accessed December 05, 2003.
2003b   Knowledge, Attitudes, and Behaviors about West Nile Virus—Connecticut,
          2002. Mortality and Morbidity Weekly Report 52(37; September
          19):886–888.
2003c   Severe Acute Respiratory Syndrome Fact Sheet. Public Health Measures in
          Response to SARS: Isolation, Quarantine, and Community Control.
          Atlanta: Centers for Disease Control and Prevention.

Chartier, Jean, and Sandra Gabler
2001   Risk Communication and Government: Theory and Application for the
          Canadian Food Inspection Agency. Nepean, Ontario: Canadian Food
          Inspection Agency, Public and Regulatory Affairs Branch.

Davis, Michael C., and C. Raj Kumar
2003   The Scars of SARS: Balancing Human Rights and Public Health Concerns.
          Public Health Practice. Electronic document, www.hk-lawyer.com/
          2003-5/May03-phprac.htm, accessed December 5, 2003.

Elliott, Susan, Mark Loeb, John Eyles, and Daniel Harrington
2003   Results of a West Nile Virus Seroprevalence Survey, South Oakville,
          Ontario, 2003. Ontario: Ministry of Health and Long-Term Care.

Epstein, Paul R.
2002   Climate Change and Infectious Disease: Stormy Weather Ahead?
          Epidemiology 13(4):373–375.

Ewald, Francois
1990   Norms, Discipline and the Law. *In* Law and the Order of Culture.
          Robert Post, ed. Pp. 138–161. Berkeley: University of California Press.

Eyles, John, and Ranu Sharma
2001   Infectious Diseases and Global Change: Threats to Human Health and
          Security. Theme issue, AVISO: An Information Bulletin on Global
          Environmental Change and Human Security 8 (June).

Feldman, Seth, Daniel Drache, and David Clifton
2003   Media Coverage of the 2003 Toronto SARS Outbreak: A Report on the Role
        of the Press in a Public Crisis. Robarts Centre for Canadian Studies.
        Electronic document, www.robarts.yorku.ca/pdf/gcf_mediacoverageSARSto
        .pdf, accessed November 29, 2003.

Finlay, Christopher
2003   The Toronto Syndrome: SARS, Risk Communication and the Flow of
        Information. Annenberg School for Communication University of
        Pennsylvania. Electronic document, from www.inter-disciplinary
        .net/transform/Finlay-Sars.pdf, accessed November 23, 2003.

Foucault, Michel
1977   Discipline and Punishment: The Birth of the Prison. London: Tavistock.

Frank, John
2003   Modernize Canada's Public Health System. Canadian Medical Association
        Journal 169(8):741. Electronic document, from
        www.cmaj.ca/cgi/eletters/169/8/741, accessed November 18, 2003.

Garland, David
2003   The Rise of Risk. In Risk and Morality. R. Ericson and A. Doyle, eds.
        Pp. 48–86. Toronto: University of Toronto Press.

Garrett, Laurie
2000   Betrayal of Trust: The Collapse of Global Health. New York: Hyperion Press.

Giddens, Anthony
1990   The Consequences of Modernity. Cambridge: Polity Press.
1991   Modernity and Self-Identity: Self and Society in the Late Modern Age.
        Cambridge: Polity Press.
1998   Risk Society: The Context of British Politics. In The Politics of Risk Society.
        J. Franklin, ed. Pp. 23–34. Cambridge: Polity Press.

Glassner, Barry
1999   The Culture of Fear. New York: Basic Books.

Goletti Delia, Drew Weissman, Robert W. Jackson, Neil M. Graham, David Vlahov,
    Robert S. Klein, Sonal S. Munsiff, Luigi Ortona, Roberto Cauda, and
    Anthony S. Fauci
1996   Effect of Mycobacterium Tuberculosis on HIV Replication: Role of
        Immune Activation. Journal of Immunology 157:1271–1278.

Guenther, Tanya, Jennifer Geduld, and Chris P. Archibald
2000   Tuberculosis among AIDS Cases in Canada from 1994 to 1999: Analysis of
        Data from the Canadian HIV/AIDS Case Reporting Surveillance System.
        Bureau of HIV/AIDS, STD and TB, Centre for Infectious Disease
        Prevention and Control, Health Canada.

Harrabin, Roger, Anna Coote, and Jessica Allen
2003   Health in the News: Risk, Reporting and Media Influence. London: King's
        Fund Publications.

Health Canada
2002a   Influenza Vaccine-Associated Adverse Events: Results of Passive
        Surveillance, Canada 2001–2002. Canada Communicable Disease Report
        28(23). Electronic document, from www.hc-sc.gc.ca/pphb-dgspsp/
        publicat/ccdr-rmtc/02vol28/dr2823ea.html, accessed November 13, 2003.
2002b   Tuberculosis: Drug Resistance in Canada, 2001. Reported Susceptibility
        Results of the Canadian Tuberculosis Laboratory Surveillance System.
        Population and Public Health Branch, Health Canada. Electronic
        document, from www.hc-sc.gc.ca/pphb-dgspsp/publicat/tbdrc01, accessed
        December 5, 2003.
2002c   West Nile Virus: Canada. Health Canada. Electronic document, from
        www.hc-sc.gc.ca/pphb-dgspsp/wnv-vwn/pdf_sr-rs/2002/situation_
        report/110102_db.pdf, accessed April 3, 2005.
2003a   An Advisory Committee Statement (ACS). National Advisory Committee
        on Immunization (NACI). Statement on Influenza Vaccination for the
        2003–2004 Season. Canada Communicable Disease Report, 29. Electronic
        document, from www.hc-sc.gc.ca/pphb-dgspsp/publicat/ccdr-rmtc/03pdf/
        acs-dcc-29-4.pdf, accessed December 3, 2003.
2003b   SARS in Canada: Anatomy of an Outbreak. Learning from SARS—Renewal
        of Public Health in Canada. Ottawa: Population and Public Health Branch,
        Health Canada.
2003c   Viruses without Borders: International Aspects of SARS. Learning from
        SARS—Renewal of Public Health in Canada. Ottawa: Population and
        Public Health Branch, Health Canada.
2003d   Tuberculosis among the Foreign-Born in Canada. Canada Communicable
        Disease Report 29(2). Electronic document, from www.hc-sc.gc.ca/
        pphb-dgspsp/publicat/ccdr-rmtc/03vol29/dr2902eb.html, accessed
        November 23, 2003.
2003e   Tuberculosis in Canada, 2001. Canadian Communicable Disease Report
        29(10):85–90.

Heywood, Neil, Barbara Kawa, Richard Long, Howard Njoo, Linda Panaro, and
    Wendy Wobeser
2003   Guidelines for the Investigation and Follow-Up of Individuals under
        Medical Surveillance for Tuberculosis after Arriving in Canada:
        A Summary. Canadian Medical Association Journal 168(12):1563–1565.

Labonte, Ronald
2003   Dying for Trade: Why Globalization Can Be Bad for Our Health. Toronto:
        Centre for Social Justice Foundation for Research and Education.

Long, Richard, Stan Houston, and Earl Hershfield
2003   Recommendations for Screening and Prevention of Tuberculosis in
        Patients with HIV and for Screening for HIV in Patients with Tuberculosis
        and their Contacts. Canadian Medical Association Journal 169(8):789–791.

MacPherson, Douglas W., and Brian D. Gushulak
2001   Migration, Population Mobility and Health: International Movement,
        Urbanization and Health. Paper prepared for the workshop "Is Your City
        Making Migrants Healthy or Sick?" at the Sixth International Metropolis
        Conference, Rotterdam, Netherlands, November 26–30, 2001. Electronic
        document, from www.international.metropolis.net/events/rotterdam/
        papers/1_MacPherson.htm, accessed November 11, 2003.

Menzies, Dick
2003   Screening Immigrants to Canada for Tuberculosis: Chest Radiography or
        Tuberculin Skin Testing? Chest Radiography versus Tuberculin Skin Testing
        for Tuberculosis Screening of Immigrants to Canada. Ottawa: Canadian
        Medical Association.

Miller, David
1999   Risk, Science and Policy: Definitional Struggles, Information Management,
        the Media and BSE. Social Science and Medicine 49:1239–1255.

Ministry of Health and Long-Term Care
2002   Organization and Delivery of Ontario's Universal Immunization Program.
        Presentation to the World Health Organization, Geneva, November.
        Electronic document, www.health.gov.on.ca/english/public/updates/
        archives/hu_02/flu/who_presentation.pdf, accessed December 8, 2003.
2003a   SARS: Severe Acute Respiratory Syndrome. Ontario: Ministry of Health
        and Long-Term Care.
2003b   West Nile Virus Preparedness and Prevention Plan for Ontario, May 27,
        2003. Ontario: Ministry of Health and Long-Term Care.

Morse, Stephen S.
1995   Factors in the Emergence of Infectious Diseases. Emerging Infectious
         Diseases 1(1). Electronic document, from www.cdc.gov/ncidod/EID/
         vol1no1/morse.htm, accessed November 22, 2003.

Nosal, Bob, and Rosnan Pellizzari
2003   West Nile Virus. Canadian Medical Association Journal 168(11):1443–1444.

Pellizzari, Rosanan
2003   Recommendations to Improve Tuberculosis Prevention and Control
         in Immigration and Refugee Residents. Toronto: Immigration and
         Refugee Working Group of the Tuberculosis (TB) Subcommittee of
         the Board of Health.

Phillips, Peter W. B.
2001   Governing Food in the 21st Century: The Globalization of Risk Analysis.
         In Governing Food: Science, Safety and Trade. Peter W. B. Phillips and
         Robert Wolfe, eds. Pp. 1–10. Montreal: McGill University Press /
         Queen's School of Policy Studies.

Sparks, Richard
2001   Degrees of Estrangement. Theoretical Criminology 5:159–176.

Tacke, Veronika
2001   BSE as an Organizational Construction: A Case Study on the Globalization
         of Risk. British Journal of Sociology 52(2):293–312.

Thacker, Stephen B., and Ruth Berkelman
1992   The History of Public Health Surveillance. In Public Health Surveillance.
         William Halperin, Edward L. Baker, and Richard R. Monson, eds. Pp. 1–15.
         New York: Van Nostrand Reinhold.

Uppaluri, Aparna, Monika Naus, Neil Heywood, James Brunton, Diane Kerbel, and
    Wendy L. Wobeser
2002   Effectiveness of the Immigration Surveillance Program for Tuberculosis in
         Ontario. Canadian Journal of Public Health 93(2):88–91.

U.S. General Accounting Office
2001   Global Health: Challenges in Improving Infectious Disease Surveillance
         Systems. Report to Congressional Requesters, GAO/NSAID 01-722.
         Washington, DC: U.S. General Accounting Office.

Venters, George A.
2001   New Variant Creutzfeldt-Jakob Disease: The Epidemic That Never Was.
         British Medical Journal 323:858–861.

Wilson, Ernest J., III
2002   Globalization, Information Technology, and Conflict in the Second and Third Worlds: A Critical Review of the Literature. Manhattan, NY: Rockefeller Brothers Fund.

World Health Organization
1999   Removing Obstacles to Healthy Development. World Health Organization Report on Infectious Disease. Electronic document, from www.who.int/infectious-disease-report/index-rpt99.html, accessed November 11, 2003.
2002   Understanding the BSE Threat. Electronic document, from www.who.int/csr/resources/publications/bse/en/BSEthreat.pdf, accessed December 5, 2003.
2003a   Summary of Probable SARS Cases with Onset of Illness from 1 November 2002 to 31 July 2003. Communicable Disease Surveillance and Response, September 26. Geneva: World Health Organization
2003b   Update 37—WHO Extends Its SARS-Related Travel Advice to Beijing and Shanxi Province in China and to Toronto, Canada. Electronic document, from www.who.int/csr/sars/archive/2003_04_23/en, accessed November 23, 2003.

Yourk, Darren, and Allison Dunfield
2003   U.S. to Lift Canadian Beef Ban. Globe and Mail, October 20.

8

Urbanization, Land Use, and Health in Baguio City, Philippines

Mary Anne Alabanza Akers and Timothy Akers

URBANIZATION AND THE DEVELOPING WORLD

Urbanization has been an ongoing process since the rise of the world's first cities, thousands of years ago. As globalization intensifies, cities are increasingly becoming the focal points for economic activity, global exchange, and disease vectors. Currently, our world is experiencing an unprecedented rate of urbanization. In 1950, only 30% of people lived in urban areas. Fifty years later, this percentage increased to 47%, and it is predicted that by 2030 more than half of the world's population will be living in urban environments (United Nations Population Division 2002:1). Although urbanization occurs in developed countries such as the United States, trends show that the fastest rates are concentrated in developing countries. In the Philippines, for example, the average annual rate of change in urban population from 1995 to 2000 is 3.64%. In 1995, approximately 54.0 million Filipinos lived in urbanized places; in a span of five years, that number increased to 58.6 million. It is estimated that in 2030, 75.1 million people will be living in urban areas. Significant migration is evident in metropolitan Manila, the national capital region; however, regional centers around the country are also experiencing this growth phenomenon. In northern Philippines, for example, the city of Baguio has a growth rate of 4.39%, which is higher than the country's urban growth rate. In 1990, the city had a population of 183,102 residents, and in 2000, its resident population hovered around 250,000, a 73.2% urban population increase over a decade.

To understand urbanization trends in the developing world, one must examine factors that lead to intense rural–urban migration. Families often abandon their farming activities when they find it difficult to compete with more productive industrial farming units. Natural disasters such as typhoons, earthquakes, and volcanic eruptions, as well as the common perception that city jobs are abundant, pull migrants to urban centers. Furthermore, the promise of better health care, education, and transportation services draws thousands of people to cities, with hopes of improving their lives. Although job opportunities and access to a range of services are greater in cities, new urban migrants are often confronted with challenges of housing shortages or slum conditions, unstable minimum wage jobs, inequitable distribution of food, unhealthy commuting situations, deteriorated water quality, potential for violence, and urban diseases. The World Bank (Porio 2002) reports that approximately 40% of Filipino urban residents live in slum and squatter communities. Furthermore, an estimated 25% of urban dwellers in metro Manila alone live in impoverished housing conditions, a much higher percentage when compared to those of other countries.

Although the detrimental effects of intense urbanization are extant in developed countries, they are far more problematic in the developing world. Environmental degradation and diminishing air, water, and land quality, for example, are all bigger problems in developing countries than in developed nations. The demands placed on natural resources far exceed the capacity that ecosystems can bear, especially in areas with high concentrations of people, machines, and structures. Changes in land uses to accommodate growing urban activities snowball to environmental problems such as river siltation, phosphorus accumulation from increased sewage, reduced air quality, and other pollution problems. People and environments collide when stressors are placed on natural resources.

As noted throughout this book, when the environment is threatened, public health is also affected. In the case presented, mercurial urban growth becomes a determining factor in predicting negative health conditions and outcomes. A report about sanitation conditions in metropolitan Manila indicates that of about two million slum dwellers, only half had access to piped water and only 17% used the services of municipal sewage systems (World Bank Group 1995). Water-borne diseases appear when urban residents rely on unsanitary sources of water. As populations continue to grow, so too does the demand for fresh water, exacerbating an already dire situation. In Southeast Asia, which has some of the world's most unsanitary urban conditions, almost one in two persons do not have access to sanitary services, and only 10% of sewage is treated at the primary level (Asian Development Bank 2001).[1] A time-series study by the Asian Disaster Preparedness Center (n.d.) revealed that malaria, dengue fever, and cholera were health consequences brought about by the combination of water shortage, climate, and humidity. In 1996, the World Health Organization (1996) reported an outbreak of cholera in Manila. An extensive study of countries around the world shows that vector-borne and water-borne diseases rose during the El Niño and La Niña periods (Hong 2000). The Philippines has been severely affected by these climactic shifts.

Malnutrition is another health problem among impoverished urban populations. The cost of food in cities is higher than that in the rural areas. Often, the male head of the household is given priority in the feeding hierarchy because his labor-intensive job requires adequate food intake for energy; since the family's source of income is dependent on his job, he must be healthy to continue working. In 1986, 46.3% of Manila's preschool children experienced first-degree malnutrition (Balanon et al. 1988). These health effects provide evidence to the negative impacts of unplanned and uncontrolled urbanization.

THE INFORMAL ECONOMY

The process of urbanization puts extreme demands on land, making it a prime commodity. Central business districts in the Philippines, for example, saw a 50% increase in land value as compared to that of peripheral areas, which grew by 25% (Porio 2002). In these conditions, people who live in poverty have more difficulty in accessing decent urban spaces for housing and livelihood activities.

Poor urban residents are resourceful; they find ways to generate income for their basic survival, such as through engaging in unregistered retail activities. They are creative and entrepreneurial, and, in spite of frequent occurrences of calamities and natural disasters, they are survivalists, thus their spontaneous participation in the informal economy. Often referred to in the development literature as the *informal sector* or *shadow economy*, this network of businesses is composed predominantly of poor women with limited access to formal labor markets. These enterprises are common in urban landscapes of any developing nation. They are found almost anywhere, inside shanty homes or out on the streets where pedestrian traffic is high; likewise, they take many forms, from a simple table stand to a small storefront. Poor urban dwellers creatively establish these unregulated business niches in the shadow landscapes of economies fraught with the inequitable distribution of resources, space, and risk realities.

Several terms describe and characterize the small unregistered businesses that operate outside the codified legalistic structures of mainstream economies—*informal, underground, hidden, black, gray.* They constitute what we call the *shadow economy* because they refer to economic activities that lack the formalistic or regulatory structures expected of "traditional" businesses (i.e., standardized systems for taxation and licensure). In this chapter, we use the terms *informal* and *shadow* interchangeably. At times, the phrase *shadow economy* can denote illegal or illicit activities, but we do not refer to that description in this chapter. In addition, we allude to street businesses as *microenterprises* to emphasize the smallness of their scale of operations. In the United States, microenterprises are defined as "very small enterprises that have, on the average, two to three workers, capital needs range from US$500 to $10,000, and are owned and operated by low to moderate income individuals" (Akers 1996:1). U.S. microenterprises are assisted by nonprofit organizations and public agencies that strongly encourage them to apply for business permits. In contrast, most microenterprises in developing countries are informal, have start-up capital of less than $100, are operated by low-income women, and are not assisted by organized groups or agencies to the extent that U.S. microenterprises are.

The shadow economy significantly contributes to the overall urban economic structure. An estimate of this sector's share in the gross domestic products of developing continents (i.e., Africa, Central and South America, Asia) is placed at 39.2% (Enste and Schneider 1998). In the Philippines, the informal sector consists of approximately 19 million Filipinos, which is over 50% of the total employment population and contributes about 44% to the country's gross national product (Asanza 2003). This significant share to the overall economic growth of the Philippines implies that further research, evaluative studies, and policy considerations should be directed to this sector. In early 2003, a conference was held in the Philippines that discussed the issue of "protecting" the informal sector in the area of health. The work of PhilHealth, a corporation owned and controlled by the government, was highlighted for its development of an innovative health insurance strategy to assist informal vendors. The impact of this initiative, however, is still to be evaluated.

Other concerns facing informal microentrepreneurs relate to substandard working conditions, lack of basic sanitation, safety, low skill levels, and lack of employment protection (e.g., disability pay). Poor women's primary roles as family caretakers and household managers, plus the responsibility of augmenting family incomes through microenterprise activities, result in added pressures.

Informal entrepreneurs in Baguio City are confronted with urban challenges typical of cities in developing nations. Microentrepreneurs visibly locate themselves in spaces with high pedestrian traffic—corners, alleyways, and sidewalks. They sell a variety of products (e.g., newspapers, candy, fresh fruit, folk herbal remedies). They keep regular hours of operations and are at the mercy of climactic conditions. Their stalls range from flat wooden boxes propped up on two- to three-feet tall cans to plastic tarps laid out on the ground. An important feature is their sharing of customers and urban space with formal businesses. This relationship between informal ventures and registered businesses can be described as both cooperative and tolerant (Akers 2001).

Responses to shadow entrepreneurs, and how well they are received, vary. Advocating for the protection of informal vendors are nongovernmental organizations such as Apalit Small Christian Communities and Lumbia Women's Self-Help Associaiton of Oro Integrated Cooperative, international organizations such as the International Labor Organization, and local groups such as the Philippine National Department of Labor and Employment. Serious discussions are taking place for initiatives to provide social protection (Asanza 2003). In general, however, the Philippine government views the informal sector negatively and attempts to control it with punitive measures. In a report conducted by the City of Baguio, the government's stance is that street and sidewalk vendors "destroy the beauty of the city and create an unpleasant scene for tourists" (City Government of Baguio n.d.:3.2). This sector was cited as one of the "unnoticed troubling issues" facing the city's economy. Earlier efforts to mitigate the shadow economy were officially passed in a city council meeting (City of Baguio City Council 1990). Complaints by pedestrians and motorists regarding vending activities that affect free flow of pedestrian traffic led to the implementation of policies governing Baguio City's informal sector.

THE RESEARCH SETTING: BAGUIO CITY, PHILIPPINES

The Philippines is a Southeast Asian country that continues to face many economic problems. In 2000, the country's total population was 76,498,735 with an average household size of five people. Although the literacy rate for the nation is more than 90%, foreign debt, political instability, and endemic fiscal ineffectiveness mark a stagnant economy. Baguio City, a regional center in northern Luzon, is 155 miles from metro Manila. Located about five thousand feet above sea level, the city has approximately 250 thousand residents, 65.5% of whom are younger than 30 years old, suggesting that it is an urban hub for higher education and medical schools and services. This mountain city is also known as one of the top travel destinations for local and international tourists. Baguio's climate is temperate, with an average of about 70 to 75 degrees Fahrenheit throughout the year. From June to October, the city receives the heaviest amount of rainfall in the country.

Baguio's initial growth is associated with international events in the early 1900s, when the United States established the city as a health resort for American soldiers and civilians. Before the 1930s, the Americans also opened mining operations in the nearby province, which led to an economic boom for Baguio. However, in World War II, the city became the target of Japanese attacks because of the concentration of Americans and their military facilities. The city was devastated by intense bombing but built itself up after the war. Reconstruction activities were swift, and Baguio became a bustling city once again. In 1990, another disaster struck. A massive earthquake rocked the city and leveled major buildings and infrastructure. Under exemplary political leadership, Baguio recovered and has continued its legacy as a primary urban center in northern Philippines. The city is also linked directly with the global economy. For example, it is home to one of Texas Instruments' international operations. In addition, many local businesses are engaged in the exportation of indigenous cultural products, such as handwoven and woodcarved items. A significant number of expatriates and foreign retirees from the United States, Australia, and Europe have settled in Baguio City, which adds an international flair to its rich cultural context.

The central business district (CBD) is in the heart of Baguio City, with major routes of transportation leading to this center. Situated on a basin-like landform, the CBD follows a grid pattern of streets and blocks with a mixture of diverse land uses. Commercial establishments, such as retail shops, restaurants, and grocery stores, are interspersed with student dorms, family apartments, and hotels. Professional offices, banks, private schools, bus stations, and the Baguio Cathedral are part of the interesting urban fabric woven throughout the CBD. The busy city market, the center to which all major streets converge, adds intensity to the district's activities. With these high levels of activity, urban space becomes a coveted prize for various users. Overlapping functions of space is evident throughout the district. For example, streets are used for unloading goods, commercial parking, ambulatory vendors, and sometimes seating for small eateries. Sidewalk spaces are shared among stationary and ambulatory vendors, registered commercial booths, parking, construction activities, and pedestrians who gather.

Baguio City's Shadow Microentrepreneurs and the Urban Environment

The present study focuses on the characteristics of informal stationary vendors in the CBD and the health challenges of working in these urban spaces. Data gathered for the study were collected in 1999 by the two of us and a team of local researchers from the University of the Philippines in Baguio.[2] A total of 211 street vendors were interviewed. Questions on business history, current business operations, and social networks were asked, while measurements of business spaces, sidewalk width, and pedestrian volume were taken to compare physical locations. As we visited these places, we repeatedly observed evidence of health and safety hazards.

Baguio City's CBD comprises six *barangays* (city districts), which we used as the basis for creating comparative zones.[3] About 35 vendor respondents were selected from each of the six barangays. Table 8.1 presents the personal characteristics of microentrepreneurs and their businesses in Baguio City. They are, to some extent, typical of street vendors in other Philippine cities and other developing nations. They are mostly women (85%) who are, on average, 43 years old and have limited education.[4] Surprisingly, however, 11.7% have college degrees.[5] The educated women have difficulty in finding formal employment in the city, so they create jobs for themselves in the informal sector, a fact consistent with the findings of a city government report (City Government of Baguio n.d.).

On average, street vendors have four children, a pattern found in another study of market vendors in the Philippines (Loanzon 1998). In our sample, at

FIGURE 8.1
Sidewalk vendors are an integral part of any urban landscape in developing communities.

least 18% of respondents had six or more dependents. Fortunately, extended relatives are integral members of Filipino households. They often help out with child care and regular household chores, leaving the female household head to focus on her microenterprise. One-third of the research sample had at least two relatives living with them at the time of the survey.

Table 8.1. Profile of "Shadow" Microentrepreneurs and Their Businesses

Characteristics	#	%
Gender		
Female	180	85
Male	31	15
Average age (in years)	43	
Educational attainment		
No formal schooling	13	6.3
Elementary	81	39.3
High school	80	38.8
Vocational	8	3.9
College	24	11.7
Household composition		
Average household size[a]	6	
Average no. of children	4	
Average no. of extended family in household[b]	2	
Female-headed households	67	32.6
Residency		
Lives in Baguio City		
Yes	194	91.9
No	17	8.1
Average no. of years lived in Baguio	29 years	
Type of business		
Fruit	51	24.1
Cooked food (in situ)	52	24.6
Newspaper/cigarettes/candy	65	30.8
Wrapped food items[c]	72	34.1
Personal accessories/toys/others	91	43.1
Average length of stay at the location	7 years	
Average daily business hours	10	
Workdays	Monday–Sunday	
Reported earnings		
Low	92	43.6
Medium	106	50.2
High	13	6.2
Reported business success		
Not successful	18	8.5
A little successful	101	47.9
Successful	88	41.7
Very successful	4	1.9

[a]Excluding respondent.
[b]In about 34% of the households.
[c]Not cooked in situ.

A study conducted in three Philippine cities indicates a rise in the proportion of female-headed households in urban centers when compared to those of rural places (Chant 1998). A significant percentage (32.6%) of the informal entrepreneurs in Baguio City were female heads of households, with no spouses living with them, suggesting that the Filipino family structure is changing. Women are becoming more economically independent and have more confidence to venture out of dysfunctional marriages. In a government report, 78% of informal vendors consider their microbusinesses as their main sources of income (City Government of Baguio n.d.).

Compared to the typical street vendor in the developing world, our respondents were not new migrants to Baguio City. A disproportionately large number of them (91.9%) lived in the city for an average of 29 years, unlike informal vendors who commuted to metro Manila daily. As a regional center, Baguio City absorbs migrants who establish themselves as permanent residents, adding extra pressure on services such as housing, transportation, water and sewer, sanitation, and health.

The informal sector in developing countries is heterogeneous and differentiated. Microentrepreneurs have diverse backgrounds and experiences and are engaged in a variety of economic activities. The informal businesses in our study are no exception. Most of them diversifed their stocks by selling an assortment of goods. Except for a few fruit vendors, respondents sold at least three different products. It is not uncommon to find a street vendor who sells wrapped food, cigarettes, and candy. Table 8.1 also shows that almost half of the vendors sell items in the "other" category, which range from posters, used books, umbrellas, and T-shirts to religious items and indigenous talismans.

Street vendors often stay in a specific spot unless compelling factors such as extremely low sales and construction activities drive them away. Even the presence of the police, who try to enforce the vending laws, does not deter them from returning to their choice spots. They simply run away until the police officer is gone, at which time they reappear to resume their businesses. In our study, vendors operated their businesses in specific locations for an average of seven years. Close to one-third of them (27.5%) had set up shop in the same spots for at least ten years. Vendors work long hours. On the average, they put in ten-hour workdays, seven days a week. Clearly, this has health implications, for they are exposed to climactic and environmental elements for long periods. These beleaguered entrepreneurs often do not realize or are indifferent to the effects of the physical environments on their health. Since vending is the main source of income for many, health dangers are overlooked.

Economically, these informal businesses are doing fairly well. About half of the vendors regarded their earnings as being moderate, while the rest thought them to be either low (43.6%) or high (6.2%). To validate their responses, we followed up with a question about business success. Table 8.2 shows that only 8.5% considered their enterprises unsuccessful. Close to 90% achieved some level of success. Initially, one may consider the results inconsistent, but the vendors viewed their microenterprises as being successful to *some* degree, despite their low earnings. Success is not viewed in monetary terms only. First, as in the case of market vendors in another Philippine city, vendors are satisfied because they know that opportunities to generate income through other means are extremely limited (Loanzon 1998). Second, good social relations with other vendors influence their perceptions of success. In our study, 68.5% of the respondents "got along very well" with other vendors and watched out for each other. For example, if a vendor were unable to work because of health problems or emergency situations, nearby vendors helped with the business until he or she returned. This form of intimate exchange has bearing on their concepts of business success. Third, success is equated with overall contentment. Many of the vendors informally admitted that they were happy with their situations and were therefore successful (Akers 1999). Nevertheless, increased income is not necessarily related to better family health and nutrition. In fact, studies have shown that when poor people generate more income, they tend to spend it on nonfood items (Rodriquez-García et al. 2001). Their nutritional conditions also deteriorate because they are apt to eat more expensive convenience, or "junk," food.

Table 8.2.   Physical Features of "Shadow Business" Locations

Zone	Vending Location (Sq. Ft.)	Typical Structure	Sidewalk Width (Ft.)	Vendors in Corners (#)	Incline of Sidewalk
1	17	Temporary	9	2	Moderate
2	21	Temporary	7	14	Low
3	12	Semipermanent	10	4	Low
4	8	Temporary	8	1	Low
5	14	Semipermanent	16	5	Low
6	15	Temporary	9	2	High

The Urban Environment and Health

When vendors are not healthy, they cannot generate the much-needed income to sustain their families. Health is an essential element for living a decent life. The urban physical environment, however, often causes the onset of a variety of health problems. Congestion, pollution, inclement weather, lack of sanitary facilities, and increased contact with people (and the contagious diseases they carry) take their toll on many urban residents. Not only is this study unique in addressing environmental health among informal entrepreneurs, but it also provides insight into the impact of microenvironments on health. The environmental health field centers on environments from a macroperspective, but we contribute to this body of knowledge by examining the physical features of small spaces and their connection to health.

Table 8.2 describes the limited spaces in which street vendors work. The average size was only 14 square feet, so they used the space creatively to display their sale products. For example, a few microentrepreneurs hung hair accessories on stall gates of formal businesses. By and large, vendors have good symbiotic relations with formal storeowners (Akers 2001). Some businesses allow vendors to keep their stocks inside their stores overnight. Table 8.2 also shows variation in vendor spaces from zone to zone. In Zone 4, for example, the average square footage for businesses was 8 square feet, while in Zone 2, vendors operated in larger spaces, averaging 21 square feet. Most street vendors used temporary structures for their businesses. Some displayed their wares on cloth tarps or baskets, while others used flat round baskets (*bilao*) on tin cans. Vendors with semipermanent stands used wooden frames or boxes, which were locked for the night and carried from one location to another if the need arose. Often, vendors sat on cans or plastic stools during working hours.

Sidewalk widths—indicative of the quality of working space—were measured throughout the study site. The average width was 10 feet, and Zone 5 had the largest sidewalk, which fronted the widest street in the CBD. We found that sidewalk measurements were proportionate to street widths. Zones 2 and 4 had narrow streets and thus smaller sidewalk widths. In terms of sidewalk incline, Zones 2, 3, 4, and 5 lay on almost-flat land, the section of the CBD that sits on a valley. Zone 6 had the steepest street slope among the six zones.

Sidewalk density was calculated by dividing pedestrian counts (number of pedestrians per minute) by sidewalk widths. The results show that Zone 3, the center of the CBD, had the highest sidewalk density of all zones (eight people per foot per minute). This is followed by Zones 6, 4, 1, 2, and 5 with sidewalk densities of 6, 5, 4, 3, and 2, respectively. The health implications of narrow sidewalk widths and heavy pedestrian traffic are several. Extreme density can lead to the spread of airborne diseases such as bronchitis, influenza, and pneumonia, the three leading diseases in the Philippines (National Statistics Coordination Board 1996). Safety is another problem associated with high sidewalk density. When sidewalks are congested, pedestrians often spill onto the streets, increasing the risk of vehicular accidents.

Another significant health predicament associated with the urban sidewalk is air pollution. An exorbitant number of respiratory problems in the developing world is of critical concern. About half of the world's 50 million to 70 million cases of respiratory illnesses are found in major cities in east Asia (McMichael 2000). The continued exposure to air contaminants such as toxic fumes from leaded-fuel vehicles, released during street vendors' long work hours, can lead to debilitating health problems. A World Bank report (Palangchao 2003) identifies Baguio City as the most polluted city in the Philippines. The report further states that the concentration of suspended particulates was found in the CBD.[6] Its topography, low wind velocity, and cold-to-moderate temperatures contribute to an inversion effect, trapping air pollution in the valley. Such highly polluted conditions have led to numerous cases of chronic bronchitis in Baguio City. In 2001, there were 311 recorded cases of chronic bronchitis, 49 of which resulted in death.

Certain zones in the study were more affected by air pollution because of the compounded effect of inversion and point-source pollution. Vendors located on two highly sloping streets (Session Road and General Luna Road) are more susceptible to health problems associated with poor air quality. Both streets are two-way, and the exhaust discharged from the bellows of vehicles moving uphill produce clouds of toxic smoke that blow directly on the vendors. Especially important to public health is the effect of pollution on young children who stay with their working mothers. Furthermore, the amount of lead particulates absorbed by food products for sale, which are typically not covered or sealed, poses a health hazard to consumers.

Water-borne diseases are another health concern. Many of the street vendors were located in flat areas that often did not have adequate drainage facilities. During the rainy season (June–November), they were exposed to stagnant water. Water runoff from upper sections brings bacteria from waste materials and street pollutants to these low-lying areas. The Asian Development Bank (2001) has reported that in Southeast Asia, more than one million people have died from diarrhea-related illnesses caused by exposure to unsanitary water conditions. Dengue, a specific health threat, is caused by poor drainage. Baguio's exorbitant rainfall increases the likelihood of stagnant water, prime breeding places for dengue-carrying mosquitoes (see Whiteford and Hill, this volume).

Street vendors are also at risk for gastrointestinal illnesses. Health officials repeatedly warn the public to be cautious about eating food sold in streets because improper handling of food can be the impetus for widespread illnesses in a community. The lack of water facilities, the presence of flies, and the unsanitary disposal of solid wastes aggravate this health problem. Public restrooms are not available in the CBD, and street vendors have limited access to private facilities for their needs.

Weather in these microenvironments pose an additional threat to vendors' health. Concrete sidewalks and walls do not protect them from extreme temperatures during the cold months of December to February, and because their businesses are not enclosed, they are also exposed to the monsoon rains. These working environments threaten their immune systems and ultimately lead to illnesses.

Gender- and age-related health issues arise as well. In a study of street vendors in South Africa, older women experienced more illnesses than did younger ones (Pick 1997). On the other hand, younger female vendors had to deal with small children who often fell sick. Regardless of age, female vendors in Baguio City no doubt experience significant stress because they engage in multiple activities to sustain their families. The long working hours and exposure to environmental health risk factors threaten their health and subsequently the ability to care for their families.

Families are also affected by the situation. Too often, older children who are left at home to fend for themselves eat less-nutritious food or engage in unsafe playing behavior. Young children who accompany their mothers suffer as well. In Mexico City, for example, a study showed that children of street vendors had a higher incidence of gastrointestinal illness and were five times more likely to suffer accidents than those who stayed home (Rodriquez-García et al. 2001). Children of Baguio street vendors, like those in Mexico City, face health risks whether they go with their mothers to work or stay at home without adequate supervision.

CONCLUSION

The study of street vendors in Baguio City, Philippines, is a good example of the relationship among urbanization, the resulting physical environment, and human health. By examining the characteristics of both macro- and micro-spaces, we extracted a number of potential health problems. However, before we suggest any policies, we need to understand the rationale behind vendors' choices of business locations. Many vendors stay in their locations for years. Ordering a transfer to healthier environments is futile because they will eventually return. When asked about location choices, at least half mentioned high customer volume. Other reasons given were familiarity with the place (30%), not being bothered by the police (17%), and less competition (16%). Only about 15% said that good shelter is the reason for their location choice.

The typical government response to street vendors is to prohibit their activities, which is neither effective nor warranted. The informal sector will always play a role in developing economies; therefore, innovative measures are essential in addressing this issue (Akers et al. 2004). A balance between supporting informal activities and protecting the mainstream economy can be achieved through creative policies. In addition, prevention-oriented health education programs are important. Respondents to the government survey claimed that spending for health-related problems is least of their priorities (City Government of Baguio n.d.). Ideally, education and awareness programs will change this attitude toward health.

Our study is part of an overarching framework of global connections. The street vendors in Baguio City are comparable to those in other developing countries. No matter how small their operations, they are affected by changes in the global exchange of goods. A shift from agriculturally based economies to manufacturing changes the nature of the economic structure and, subsequently, the relationships among people, the means of production, and the land. Where once people were self-sufficient, they have become dependent on industrial low-paying jobs. Members of the household are forced to migrate to urban centers and engage in informal business activities that have numerous health risks. When governments make economic decisions that have global implications, marginalized populations are most affected. Policymakers need to understand the complex web of connections among people and social structures around the world and consider the local ramifications of their decisions.

NOTES

We wish to acknowledge the Cordillera Studies Center at the University of the Philippines, Baguio, for facilitating the hiring of competent research interviewers and for the use of its facilities during the data-gathering phase.

1. "Primary level" means a type of treatment process that separates solid sewage from liquid sewage in a settling tank. This implies a modern sewage system that many developing countries yet do not use at a large scale.

2. We completed a follow-up study in December 2003. While its data are not presented in this chapter, the findings are consistent. However, in this second study, we focus more extensively on actual health conditions of vendors in Baguio City, Philippines. The findings are forthcoming in other publications.

3. A *barangay* is the smallest yet most significant geopolitical unit in the Philippines. Historically, *barangays* were community villages of about 50 to 100 families but have grown now to encompass several thousand. It is within this granular structure that voices of people are heard and government programs are implemented.

4. A high school education in the Philippines is considered low educational attainment.

5. A statistically significant relationship was found between education and gender (Chi square $= 21.831$, $p < .05$). Of those with college degrees, 19 out of the 24 vendors were women.

6. The Total Suspended Particle (TSP) level in Baguio's central business district exceeds the Philippine Department of Environment and Natural Resources' standard value of 90 micrograms per cubic meter.

REFERENCES

Akers, Mary Anne Alabanza
1996   Flexible Manufacturing Networks for Microenterprises. Washington, DC: National Congress for Community and Economic Development.
1999   The Use of Public Spaces for Economic Gain: A Study of Informal Street Businesses in the Philippines. Paper presented at the 31st Environmental Design Research Association Conference. San Francisco, California, May 10–14.
2001   Relationships between Formal and Informal Businesses. Poster exhibit at the 32nd Environmental Design Research Association Conference. Edinburgh, Scotland.

Akers, Mary Anne Alabanza, Richard Sowell, and Timothy Akers
2004   A Conceptual Model for Planning and Designing Healthy Urban
        Landscapes in the Third World: A Study of Street Vendors in the
        Philippines. Landscape Review 9(1):45–49.

Asanza, Anna Lucila
2003   Extension of Social Protection in the Philippines. Manila: International
        Labour Organization. Electronic document, www.ilo.org/public/english/
        region/asro/manila/2003/mar/espp.htm, accessed April 25, 2003.

Asian Development Bank
2001   Asian Environmental Outlook. Publication Stock, 020600. Manila,
        Philippines: Asian Development Bank.

Asian Disaster Preparedness Center
N.d.   Expected Impacts of El Niño in Indonesia, Philippines, Vietnam. Electronic
        document, www.adpc.net/ece/NewsEvents/ElninoImpacts.html, accessed
        March 27, 2005.

Balanon, Lourdes, Delia Jimenea, Chona Bravante, and Leopoldo M. Moselina
1988   The Situation of Street Children in Ten Cities. Manila, Philippines:
        Department of Social Welfare and Development, National Council of
        Social Development Foundation of the Philippines, and the United Nations
        Children's Fund.

Chant, Sylvia
1998   Households, Gender, and Rural-Urban Migration: Reflections on Linkages
        and Recommendations for Policy. Environment and Urbanization 10(1):5–21.

City Government of Baguio
N.d.   A Study of the Underground Economy. Baguio, Philippines: City Government
        of Baguio, Office of the City Planning and Development Coordinator.

City of Baguio City Council
1990   City Council Minutes, January 24.

Enste, Dominik, and Friedrich Schneider
1998   Increasing Shadow Economies All over the World—Fact or Reality.
        Discussion Paper, 26. Bonn, Germany: Institute for the Study of Labor (IZA).

Hay, Maureen
1987   The Informal Sector in Third World Cities: A Geographical Analysis of
        Street Vending in Peru. Discussion Paper, 91. Syracuse, NY: Department of
        Geography, Syracuse University.

Hoefer, Johannes Hans
1990   Philippines. Hong Kong: APA Publications.

Hong, Evelyne
2000   Globalization and the Impact on Health: A Third World View. Issue Paper
         prepared for the People's Health Assembly, Svar, Bangladesh. Electronic
         document, http://phmovement.org/pdf/pubs/phm-pubs-hong.pdf, accessed
         March 27, 2005.

Loanzon, Jeanette Isabelle
1998   Hanap-Buhay: Securing Economic Contributions of Poor Woken through
         an Innovative, Community-based, Self-Help, Non-Governmental
         Development Organization cum Cooperative—a Case Study of the Market
         Women of Apalit, Pampanga, Central Luzon, Philippines. Manila:
         University of Santo Tomas Publishing House.

McMichael, Anthony, Jr.
2000   The Urban Environment and Health in a World of Increasing
         Globalization: Issues for Developing Countries. Bulletin of the World
         Health Organization 78(9):1117–1123.

National Statistical Coordination Board
1996   Vital, Health, and Nutrition Statistics: Sectoral Statistics. Manila: National
         Statistical Coordination Board.

Palangchao, Harvey
2003   Baguio Officials Jolted into Action on Dirty Air Report. Manila Times,
         March 18.

Pick, William
1997   Household-Related Variables and Reported Illness in Street Vendors and
         Their Children in a South African City. Working Paper, RP134. Cambridge,
         MA: Harvard University School of Public Health / Takemi Program in
         International Health.

Porio, Emma
2002   Urban Poor Communities in State-Civil Society Dynamics: Constraints
         and Possibilities for Housing and Security of Tenure in Metro Manila.
         Asian Journal of Social Sciences 30(1):73–96.

Rodriquez-García, Rosalia, James A. Macinko, and William F. Waters
2001   Microenterprise Development for Better Health Outcomes. Westport, CT:
         Greenwood Press.

United Nations Population Division

2002   World Urbanization Prospects: 2001 Revisions. Electronic document, www.un.org/esa/population/publications/wup2001/WUP2001_CH1.pdf, accessed March 27, 2005.

World Bank Group

1995   Sewerage and Sanitation: Jakarta and Manila. Precis 91(June). Electronic document, http://wbln0018.worldbank.org/oed/oeddoclib.nsf/0/ 4BE7A12A7DD3B01A852567F5005D897C?OpenDocument, accessed March 27, 2005.

World Health Organization

1996   Cholera in the Philippines. Disease Outbreaks Reports, September 12.

9

Globalization, Demography, and Nutrition: A Bekaa Bedouin Case Study

Suzanne E. Joseph

The goal of this chapter is to provide an integrated perspective of demographic behavior and nutritional status among Bedouin agropastoral tribes in the Bekaa Valley, Lebanon. In doing so, I hope to illustrate the relationship between globalization—manifest in a greater reliance on agriculture and integration into a market economy—and fertility, mortality, and nutrition at the local population level. Just as the history of disease and globalization are strongly interrelated, demographic processes and globalization are linked in time and space.

George J. Armelagos and Kristin N. Harper (this volume) describe how sedentary life and population growth are historically associated with agricultural development. The Bedouin case study demonstrates that, in the short term, greater reliance on agriculture is associated with fertility increase at the local level. However, in addition to the fertility increase that has accompanied the peasantization of the Bedouin social economy, we see low mortality and adequate nutrition. While evidence of good Bedouin health may appear to contradict the literature reviewed by Armelagos and Harper, which reveals a decline in nutrition and increase in infectious disease with agricultural development, it is important to remember that in the contemporary context, Bedouins are experiencing two transitions simultaneously. That is, in addition to fertility changes associated with increased dependence on agriculture and sedentary life, mortality changes associated with Bedouin access to modern medical care are also apparent. The decrease in infectious diseases and the subsequent reduction in infant mortality have resulted in greater life expectancy at birth for Bedouins.

From a demographic and health perspective, the Bekaa Bedouin may thus be characterized as a well-nourished, high-fertility, low-mortality population that is currently undergoing fertility decline. Historically, greater Bedouin reliance on agriculture can be linked to the replacement of camels with pickup trucks, which is in turn tied to broader political–economic developments, including the delineation of political boundaries, alienation of tribal lands by governments, population growth, and agricultural expansion. Such political economic changes are, in the short term, associated with high fertility and good health. However, over the long term, greater integration into a cash economy appears to have triggered Bedouin fertility decline. With monetization, families become increasingly vulnerable to the vagaries of the market. The downward trend of the national economy since the late 1980s, seen in the depreciation of the national currency, galloping inflation, and the decline in real wages, seems to have forced Bekaa Bedouin fertility downward.[1] Hence, fertility decline among this group appears to be linked to increasing economic hardship, further evidenced at the local level in the rising proportion of wage laborers in the Bedouin population.

The chapter begins with a summary of research methods used to determine subsistence activity, demographic outcomes, and nutritional status. The second section provides a historical overview of Bedouin subsistence practices. The third section presents research findings on Bedouin demographic and nutritional profiles in the context of an increasingly agricultural economy. The discussion ends with a summary of research findings, and it highlights the future challenges facing Bedouin family formation and nutritional status with their integration into a global economy.

RESEARCH METHODS

The demographic, nutritional, and economic subsistence data presented here are based on a systematic random sample of 240 ever-married Bedouin women between the ages of 15 and 54 in the Bekaa Valley, Lebanon.[2] Research was conducted among agropastoral Bedouin tribes (of which there are over seven) in the Bekaa Valley for a period of 15 months in 2000–2001. The Bekaa Valley is an upland valley located between the Lebanon Mountains and the

Anti-Lebanon Mountains. Bedouin women between the ages of 15 and 54 were selected from every sixth household in Bedouin villages, hamlets, and neighborhoods in the Bekaa.[3] The Bekaa is about 177 kilometers in length and 10 to 16 kilometers wide with an average elevation of 762 meters.

Demographic measures were derived retrospectively from women's reproductive histories (Howell 1979, 2000). In reproductive histories, every offspring is listed by birth order, and information about the year of birth and year of death, as well as the age of the mother, is recorded. Because some of the older women had difficulty allocating their exact ages and the dates of events, I created a local event calendar to reduce age-estimation error (Freedman et al. 1988).

Anthropometrics are used to evaluate the nutritional status of Bekaa Bedouin women. Information about the nutritional status was derived from measurements of height, weight, skinfold thicknesses at the triceps and subscapula, and mid–upper arm circumferences. All anthropometric measurements were taken following the protocols of T. G. Lohman and colleagues (1988).

Information about subsistence or production mode was derived from work histories. Each household was categorized according to its primary mode of production since a couple's marriage. Oral histories and ethnographic methods were also used to understand the cultural and historical context of social relations of production.

ECONOMIC SUBSISTENCE BASE OF THE BEDOUIN: A HISTORICAL OVERVIEW

Nomadic pastoralism refers to the mobile or semimobile herding of livestock in arid and semiarid lands. Pastoralists rely on the products (e.g., meat, animal fibers, animal skins, and dairy products) of their herds (e.g., sheep, goats, camels, cows, reindeer, and horses), which they consume, barter, or sell. The extent to which pastoralists rely on nonpastoral (particularly agricultural) products varies cross-culturally (Dyson-Hudson and Dyson-Hudson 1980).

Traditional nomadic pastoralism among the Bekaa Bedouin has been altered by processes similar to those affecting pastoralists everywhere. Bedouin pastoral movements in search of pasture and water for their sheep and goats have been drastically altered by national frontiers and policies, changes in systems of land tenure, mechanization of transport, population growth, and by the expansion of cultivation.

The Bekaa Valley is central to the traditional pastoral cycle of Bedouin tribes. Historically, households migrated as a unit from winter to spring pastures using camels to transport their tents and possessions. Traditional winter grazing lands were east of Homs (in the direction of Tudmor or Palmyra) in the Syrian Desert (*Badiyat Ash Sham*; Chatty 1974). The Syrian Desert is an upland plateau scored with numerous wadis and covered with grass and scrub vegetation, which are used extensively for pasture by both nomadic and semi-nomadic herders. These traditional winter pastures are in bloom in early January, February, and March. The lambing of the flock occurs in February or March.

Traditionally, tribes that did not migrate to these winter pastures usually headed to the northern Bekaa caves of al-Qaa to shelter their sheep from the cold and wet winter. During April and early May, tribes began moving back toward the Bekaa. They camped in scattered and remote areas along the Anti-Lebanon Mountains, where sheep could graze on natural pastures (Chatty 1974). During the spring–summer season, Bedouin families made butter from sheep milk, which was then sold to middle men with whom they came into contact in their southward movement toward the Bekaa.

Toward the close of the spring season, arrangements were made between household heads and peasant landowning farmers whereby the sheep could graze on the stubble of harvested fields and in turn fertilize the farmer's fields with manure (Chatty 1974). Such arrangements were considered mutually beneficial. In oral history interviews, Bedouin households reported that seasonal relationships with peasant landowners were maintained for long periods, often lasting over ten years (Joseph 2002). This means that Bedouin pastoral families did not have to renegotiate oral agreements with peasant landowners season after season. Pastoralists remained in the Bekaa until the winter season (late September or early October), which again marked their movement north.

A fundamental change to Bedouin pastoralism occurred in the 1960s, as camels, which had been used for transport, were replaced with pickup trucks (see Chatty 1986). The mechanization of transport led to major reductions in the nomadic movements of Bedouins and greater peasantization and commercialization of their social economy. First, trucks drastically reduced the mobility of pastoralists. Although the truck does restrict camping grounds to areas that can be reached by roads, its use released camping units from the marched migrations, which were several weeks long (between winter and spring pastures). With the truck, the same journey was possible in as little as three hours. Oral histories indicate that before the 1960s, the Bedouin moved at least seven times during the spring and summer periods of the pastoral cycle (Joseph 2002).

As families began to remain in the Bekaa Valley for longer periods, they developed stronger ties to peasant society, hiring out their trucks and services to neighboring villages and towns (Chatty 1974:121). By the 1970s Bedouin household production shifted from the exchange of butter to the sale of sheep and goat milk to dairy companies. The work roles of young Bedouin women also changed. That is, by 1972–1973, new agreements were being made in which young unmarried Bedouin girls were collectively transported by pickup trucks (belonging to either peasant landowners or Bedouin pastoralists) to work in the farmer's fields and returned to their residential camping units collectively (125). Unmarried Bedouin women normally work in groups alongside kin. Such work arrangements help circumvent cultural concerns over female honor and chastity.

D. Chatty (1974:152) states that by 1972–1973, tribal units and their herds entered the Bekaa as early as March and remained until late November. Oral histories from the present study similarly indicate that many households during the 1970s ceased their seasonal migratory movements to the Syrian Desert and began purchasing land in the Bekaa for household construction. Most land was jointly purchased by brothers or patrilateral parallel cousins. Although Bedouins purchased land in the mid- to late 1960s, most did not begin building houses until the 1970s and 1980s. To cover the costs of land purchase and household construction, Bedouin families began selling a substantial portion of their livestock (and, in some cases, all). The sale of livestock meant that tribes came to rely less on pastoralism and more and more on agricultural sharecropping and wage labor.

In terms of the current economic structure of the sample population, most Bedouin households rely on agricultural sharecropping (99/240) for their livelihoods, while roughly one-fourth rely on wage labor (65/240) and pastoralism (64/240). The remaining Bedouin households (12/240) are involved in miscellaneous labor pursuits, mostly small shopkeeping or petty trade.

Since Bedouins as a group are land-poor (i.e., 71% of families own only the land on which their house rests), farming households are involved in oral sharecropping agreements with Lebanese landowning peasants in the Bekaa.

The type of sharecropping (*sheerak*) among Bedouin households can be defined as a shared investment by the landlord and tenant farmer in which the landlord provides the land and water, half of all the other inputs (seedlings, fertilizers, pesticides), while labor (harvesting, weeding, and so on) is solely provided by the tenant (see Pollock 1989). Machinery is sometimes provided by the landlord and sometimes rented. Each party receives half the net crop yield. In sharecropping, legal ownership of the land resides not in the hands of direct producers but in the hands of nonlaborers—in this case, Lebanese landowning peasants. While there are a few Bedouin families who rely on tenant farming, sharecropping is much more common.

Most Bedouin wage laborers in the Bekaa work in agriculture (and are usually paid a daily wage), although some are employed in industry, primarily local agroindustries, such as chicken and cheese factories. Bedouin pastoralism also involves close ties to peasant agriculture. Bedouin pastoralists are involved in a modern semisedentary form of shepherding. During the period of fieldwork in 2000, Bedouin pastoralists were migrating twice on average—once in the spring, when they head for the natural grasses of the Lebanon and Anti-Lebanon mountains, and once in the summer, when they rent farmland from peasants in the Bekaa Valley for grazing. During the winter, most pastoralists rely on purchased feedstuffs. To protect sheep and goats from the cold winter months, families house their flocks in makeshift enclosures.

In sum, the major economic trend over the last four or so decades has been the decline and modernization of pastoralism, as well as a greater reliance on farming and wage labor. While most Bedouin families in the Bekaa rely on farming for their livelihood, both pastoral and farming families depend on the income derived from wage labor of women and children. Hence, the proletarianization of labor is an important feature of the Bedouin social economy.

DEMOGAPHY AND NUTRITION

Fertility

The total fertility rate is used to characterize the overall level of Bedouin fertility, since it provides a way to interpret measure of reproductive performance. The total fertility rate is defined as the expected number of offspring ever born to a randomly selected woman who survives to the end of the reproductive span (i.e., to menopause or to an advanced reproductive age).

The total fertility rate can be calculated as a cross-sectional measure by summing the population's current age-specific fertility rates. Another way of calculating the total fertility rate is to take the mean number of live births ever born to women of postreproductive age—a retrospective measure of cohort fertility. However, K. L. Campbell and J. W. Wood (1988) have cautioned that if mortality is selective with respect to fertility (i.e., if reproduction itself puts women at a greater risk of death), then retrospective total fertility rates may be biased downward, simply because women with lower fertility have a better chance of surviving long enough to be included in the sample.

It is important to remember that period and cohort total fertility rates are expected to be the same only if fertility rates have remained stable over time. The total period fertility rate of ever-married Bedouin women aged 15 to 49 is 5.5. However, the total fertility rate calculated from the children ever born to a smaller subsample of women postreproductive age or near postreproductive age is 8.9 (Joseph 2004). The fact that the total cohort measure of fertility (8.9) does not coincide with the total period measure of fertility (5.5) is indicative of recent fertility decline in the younger population.

While there is evidence of recent fertility decline, the mean parity of pretransitional Bekaa Bedouin women is nearly as high as that of the Hutterites (see Joseph 2004). High fertility among the Bekaa Bedouin can be linked to economic shifts that include the mechanization of transport, villagization, and greater reliance on agriculture. A growing body of research suggests the presence of a significant average difference in fertility between agriculturalists and nonagriculturalists across human populations (Bentley et al. 1993; Sellen and Mace 1997). It is important to remember that analyses performed at such highly aggregated levels may contribute little insight to the causes of fertility variation at regional and community levels or among individuals within a local population (see Campbell and Wood 1988). Aggregate cross-sectional data also provide little understanding of microhistorical processes of change.

A more complete understanding of the causes behind high Bedouin fertility can be achieved by focusing attention on the proximate determinants or mechanisms of fertility. The proximate determinants of fertility refer to those biological and behavioral mechanisms that directly impinge on the number of live births. The logic behind the proximate determinants framework is that, because biological reproduction is a regular process, it is possible to come up with a short list of variables that have a direct and inevitable effect on fertility—inevitable in the sense that any change in one of these variables translates into a change in fertility. The proximate determinants framework is not meant to be reductionistic (James Wood, personal communication, January 2002) but to allow researchers to hierarchically link different features of behavior. That is, finer-level mechanisms explicate the "hows" of behavior, whereas broader levels explain the "whys" of behavior (see Allen and Hoekstra 1992).

The proximate biological and behavioral determinants of fertility are postpartum–lactational infecundability, or the mean duration of the infertile period following birth; the waiting time to conception, which depends on fecundity and coital frequency; intrauterine mortality, or the number of pregnancies that prematurely end in miscarriages and stillbirths; permanent sterility, or the proportion of men and women in sexual–marital unions who are sterile in the population; entry into the reproductive span, which is a function of age at menarche and age at marriage; the effectiveness of contraception; and the rate of abortion (Bongaarts and Potter 1983; see also, Wood 1990).

Among the Bedouin, high fertility is linked to early weaning, early and stable marriages, and low levels of primary sterility (Joseph 2004). Bedouin women have relatively short median birth intervals (25 months) due to early weaning. On average, women marry at the age of 18, and their marriages are highly stable. Marital stability in Bedouin society can, in part, be attributed to strong kinship ties, endogamous marriage, and possession of a distinct cultural identity—distinct in some key aspects from that of the dominant culture (see Joseph 2002:78–114). Of additional importance in understanding high Bekaa Bedouin fertility is the low incidence of pathological sterility. Only one woman in the subsample of 65 women aged 40 and over reported zero parity, which indicates that roughly 1.5% of married postreproductive Bedouin women are unable to conceive children. This is not an unexpected observation since the reproductive system is sufficiently complex and has a certain probability of failure. Empirical findings in other noncontracepting populations indicate that some 3–10% of married postreproductive women have zero children (Howell 2000:125).

Findings presented elsewhere indicate that high fertility among the Bedouin was of brief historical duration, characterizing women born between 1934 and 1960 (Joseph 2004). These women were marrying and starting their families during a period of fundamental economic transition that coincided with the mechanization of transport and greater peasantization of the

Bedouin economy. Although future demographic research is needed to better understand the causes of recent fertility decline in Bedouin society, it appears that families are facing increasing economic hardship in the context of an economic recession in Lebanon.

Mortality

Lifetable estimates of Bekaa Bedouin infant mortality (0.05) and child mortality (0.01) point to relatively low mortality in comparison to those found in other traditional non-Western societies prior to fertility transition (Pennington 1996). In examining available data on the mortality experiences of 13 small-scale "preindustrial" non-Western populations, R. Pennington (1996:263) finds that infant mortality ranges from a low of 0.06 among recently born Herero of Botswana to a high of 0.24 among ethnically mixed Mandinka and Jola populations in the Gambia. Child mortality rates range from a low of 0.03 among the Herero of Botswana to a high of 0.36 among Delta Fulani children in Mali.

The relatively low mortality experiences of the Bedouin are in large measure due to the widespread availability of medical and health care facilities in the region. The availability of high-quality medical and health care services as well as adequate public sanitation in the Bekaa substantially minimizes the risk of child death. In terms of maternal delivery experiences, virtually all Bedouin women deliver their babies in local clinics with the assistance of well-trained peasant midwives. While some older women did report delivering their first child at home, in most cases a midwife or older experienced Bedouin woman was present to provide delivery assistance, particularly in the removal of the placenta.

Low Bedouin infant and child mortality rates may also be linked to the presence of high-quality weaning foods, particularly sheep and goat milk, in Bedouin society (see Pennington 1996). Aggregated analyses performed by D. Sellen and R. Mace (1999)—using the Human Relations Area Files and other anthropological sources—indicate a positive association between dependence on extractive modes of subsistence (hunting, gathering, and fishing) and total child mortality. Their data suggest that a 10% increase in dependence on extracted resources is associated with an increase in childhood mortality of just under 30 deaths per 1,000 (8). However, cross-cultural comparisons indicate that infant mortality rates do not vary with subsistence activity. Hence, the authors conclude that it is

problematic to argue that the availability of easily digestible nutrient-rich weaning foods (i.e., dairy or cereal products) in pastoral and agricultural economies
substantially reduces infant mortality. The authors argue that the protective effects of continued breast-feeding throughout the first year of life—believed to be
omnipresent in traditional societies—served to reduce the variance in infant
mortality across subsistence economies.

Nutrition

Information about the nutritional status of women in the sample was derived
from measurements of height, weight, and skinfold thicknesses at the triceps and
subscapula and from mid–upper arm circumference. The anthropometric characteristics of Bedouin women are given in Table 9.1. Relative to the published
standards of A. R. Frisancho (1990), Bedouin women fall between the 10th and
15th percentiles for stature. Small stature is believed to be indicative of much
longer-term nutritional deprivation or dietary insufficiency. However, in terms
of body mass index (weight/height^2), subcutaneous fat stores, and protein-
energy stores, the Bedouin are considered well nourished.

A good indicator of adult health is one's weight in proportion to one's
height (weight for height), or body mass index (BMI). Bedouin women fall
between the 75th and 85th percentiles for BMI. According to the Centers for
Disease Control and Prevention (2004), the Bekaa Bedouin are in the over-
weight range (25.0–29.9). However, it is important to keep in mind that excess
weight does not necessarily imply excess fat. A large-framed muscular adult
may have a high weight for height. In terms of the sum of triceps and sub-
scapular skinfold thicknesses, the Bedouin fall between the 50th and 75th percentiles for their age. For upper-arm muscle area, Bekaa Bedouin women fall
between the 75th and 85th percentiles. When using skinfold thicknesses and

Table 9.1.   Anthropometric Characteristics of Bekaa Bedouin Women

Variable	M	SD
Age	33.9	9.2
Weight (kg)	64.4	14.0
Height (cm)	155.2	5.7
Body mass index (kg/m^2)	26.7	5.7
Triceps skinfold thickness (mm)	24.2	14.3
Subscapular skinfold thickness (mm)	24.4	11.1
Arm muscle area (cm^2)	35.3	8.1

Note: M = mean, SD = standard deviation.

other anthropometric measurements of fat status (i.e., mid-arm fat area and mid-arm fat index; see Frisancho 1990), the Bekaa Bedouin fall in the upper end of the average range (i.e., 15th to 75th percentile).

Table 9.2 provides mean height, weight, and BMI for Bekaa Bedouin women and women from three Middle Eastern countries for comparison. Table 9.2 reveals that Bekaa Bedouin women are comparable in height, weight, and BMI to other Middle Eastern women, with Jordanian women having the highest BMI and Yemeni women the lowest BMI. Hence, according to the anthropometric measurements presented here, it is clear that Bedouin women's nutritional status is not below regional Middle Eastern standards.

Recent studies of Bedouin children's food consumption and nutritional status in different parts of Lebanon reveal that nutritional deprivation was more pronounced among settled Bedouins who received rations from the government than it was for Bedouins who had a diversified subsistence base (Baba et al. 1994). Bekaa Bedouin children were found to have better nutritional status than that of Bedouin children in other parts of Lebanon. The higher nutritional status of Bekaa Bedouin children was attributed to income derived from small-scale cash crop production, livestock milk production, and wage labor.

Supplementary income derived from wage labor was afforded in large part by massive rural to urban migration. M. Daher describes how many villages in rural Lebanon were emptied of their young workforce as a result of both internal migration and emigration to foreign countries (1992:44). Such migration appears to have left a void in the rural economy, which Bekaa Bedouins, particularly adolescent girls, were able to fill (Joseph 2002). Virtually all Bekaa

Table 9.2.   Bekaa Bedouin Women Compared with Women of Three Middle Eastern Countries

	Weight (kg)		Height (cm)		Body mass index (kg/m²)	
	M	SD	M	SD	M	SD
Bekaa Valley, Lebanon	64.3	13.9	155.2	5.7	26.7	5.7
Jordan	67.8	14.3	157.5	5.7	27.3	5.5
Turkey	63.3	12.5	156.2	5.7	25.8	4.9
Yemen	49.9	10.4	152.7	6.6	21.2	4.2

Source: For the three Middle Eastern countries, Demographic and Health Surveys online (www.measuredhs.com). Data for Jordan (1997), Turkey (1998), Yemen (1997).
Note: M = mean, SD = standard deviation.

Bedouin households, regardless of their dominant form of production, rely on the extra income derived from seasonal wage labor. Hence, a mixed agropastoral economy is key to understanding the demographic and nutritional profiles of the Bekaa Bedouin.

CONCLUSION

Pastoralist societies throughout the world are facing increased threats to their livelihoods, with increased commoditization of their livestock economy and labor, state and private encroachments on tribal lands, sedentarization, and population growth (Fratkin 1997). Pastoral populations continue to herd their animals in arid and semiarid lands of the Middle East, Africa, Central Asia, Mongolia, highland Tibet and the Andes, and arctic Scandinavia and Siberia. Today, encapsulation continues as industrial states expand commercial production in arid lands, forcing pastoralists into sedentary or urban communities (Khazanov 1994). In the Middle East, pastoralists are increasingly becoming incorporated into agrarian peasant state societies under the authority of a politically and economically dominant ruling class (Salzman 1999). Addressing the sociodemographic and health consequences of political encapsulation and reduced pastoral autonomy and mobility is critical to understanding the challenges pastoralists are facing in the modern world.

Numerous studies point to the detrimental social and economic impacts of pastoral sedentarization and commoditization, including poorer nutrition and a lack of adequate housing and drinking water (Fratkin 1997). However, sedentary groups seem to have access to better health care. In the Bekaa Bedouin setting, high maternal and child nutritional status and moderate–low mortality are linked to a flexible or mixed agropastoral economy with access to modern health care. At this historical juncture, Bedouins can be characterized as a well-nourished population. In terms of child survival, infant and child mortality are low in comparison to that of other non-Western "preindustrial" societies. Low mortality is indicative of high-quality medical care facilities in the Bekaa, dietary sufficiency, and low disease loads. Cohort measures of fertility among postreproductive women indicate that the Bedouin have one of the highest fertility rates ever observed, although there is evidence of recent fertility decline among younger women. High fertility is found in the context of a mixed economy with a heavy reliance on agriculture.

However, the historical decline of pastoralism in Bedouin society is cause for concern since a mixed economy allows for diversification of income, which appears to provide a nutritional advantage. As more and more families relinquish their herds (to buy land or build a house) and lose a fundamental part of their economic livelihoods, a future decline in their overall health is likely to occur. Prospective research is necessary to determine whether or not greater integration of Bedouins into a global economy will result in poorer nutrition.

Vital demographic events have also been affected by political–ecological change. With villagization and sedentarization, Bedouin marriage and family formation are increasingly tied to household construction, which entails considerable expense. Similarly, shepherding increasingly requires money, which is needed to purchase vitamins, vaccines, and winter feedstuffs and to rent farmland from peasants during the summer and spring grazing months. That is, Bedouin livestock owners pay per *dunum* (1 *dunum* equals one quarter of an acre) to graze their flocks on the stubble left after crops are harvested (the price of harvest yieldings is determined by harvest crop). Trucks also require repair, maintenance, and fuel—which only add to the expenses now facing Bedouin households. The challenge of meeting such household expenses is exacerbated during times of economic recession.

Hence, Bedouin families are confronting a different social and economic order, one that increasingly involves weighing expenses and coming to terms with growing demands on the household exchequer. In terms of fertility behavior, Bedouin families already appear to have adjusted to these new constraints in their sociocultural environment by controlling their fertility within marriage. Younger couples express a clear desire to limit their family size, and many are turning to modern contraception to do so.

NOTES

1. This period marked a major turning point in the Lebanese economy and was triggered, in part, by the Israeli invasion of 1982 and by declining remittances from the Gulf due to the drop in oil prices.

2. Upon rechecking, 1 of the 240 Bedouin women in the sample was found to be 56 years old but was still included in the demographic analyses of postreproductive women.

3. Bedouins living in shacks and tents in small, more temporary enclaves were also included in the sample.

REFERENCES

Allen, T. F. H., and T. W. Hoekstra
1992   Toward a Unified Ecology. New York: Columbia University Press.

Baba, N. H., Shaar, K., Hamadeh S., and N. Adra
1994   Nutritional Status of Bedouin Children Aged 6–10 Years in Lebanon and
          Syria under Different Nomadic Pastoral Systems. Ecology of Food and
          Nutrition 32:247–259.

Bentley, G. R., G. Jasienska, and A. Goldberg
1993   Is the Fertility of Agriculturalists Higher Than That of Non-Agriculturalists?
          Current Anthropology 35:778–785.

Bongaarts, J., and R. G. Potter
1983   Fertility, Biology, and Behavior: An Analysis of the Proximate
          Determinants. New York: Academic Press.

Campbell, K. L., and J. W. Wood
1988   Fertility in Traditional Societies. In Natural Human Fertility: Social and
          Biological Determinants. Edited by P. Diggory, M. Potts, and S. Teper,
          Pp. 39–69. London: Macmillan.

Centers for Disease Control and Prevention
2004   What Is BMI? Electronic document, www.cdc.gov/nccdphp/dnpa//bmi/
          bmi-adult.htm, accessed January 14, 2004.

Chatty, D.
1974   From Camel to Truck: A Study of the Pastoral Economy of the Al-Fadl and
          Al-Hassana in the Bekaa Valley, Lebanon. Ph.D. dissertation, Department
          of Anthropology, University of California, Los Angeles.
1986   From Camel to Truck: Bedouin in the Modern World. New York:
          Vantage Press.

Daher, M.
1992   The Socio-Economic Changes and Civil War in Lebanon 1943–1990.
          Tokyo: Institute of Developing Economies.

Dyson-Hudson, R., and N. Dyson-Hudson
1980   Nomadic Pastoralism. Annual Review of Anthropology 9:15–61.

Fratkin, E.
1997   Pastoralism: Governance and Development Issues. Annual Review of
          Anthropology 26:235–261.

Freedman, D. A., D. Thornton, D. A. Camburn, and L. Young-Demarco
1988   The Life History Calendar: A Technique for Collecting Retrospective Data.
        In Sociological Methodology. C. C. Clogg, ed. Pp. 37–68. Washington, DC:
        American Sociological Association.

Frisancho, A. R.
1990   Anthropometric Standards for the Assessment of Growth and Nutritional
        Status. Ann Arbor: University of Michigan Press.

Howell, N.
1979   The Demography of the Dobe !Kung. New York: Academic Press.
2000   The Demography of the Dobe !Kung. 2nd edition. New York: Aldine
        De Gruyter.

Joseph, S.
2002   Forms of Production and Demographic Regimes: An Anthropological
        Demographic Study of Bedouin Agro-Pastoral Tribes in the Bekaa Valley,
        Lebanon. Ph.D. dissertation, Department of Anthropology, University
        of Georgia.
2004   The Biocultural Context of Very High Fertility among the Bekaa Bedouin.
        American Anthropologist 106(1):140–144.

Khazanov, A. M.
1994   Nomads and the Outside World. Madison: University of Wisconsin Press.

Lohman, T. G., A. F. Roche, and R. Martorell
1988   Anthropometric Standardization Reference Manual. Champaign, IL:
        Human Kinetics.

Pennington, R.
1996   Causes of Early Human Population Growth. American Journal of Physical
        Anthropology 99:259–274.

Pollock, A.
1989   Sharecropping in the North Jordan Valley: Social Relations of Production and
        Reproduction. *In* The Rural Middle East: Peasant Lives and Modes of
        Production. K. Glavanis and P. Glavanis, eds. Pp. 95–121. London: Zed Books.

Salzman, P.
1999   Is Inequality Universal? Current Anthropology 40(1):31–61.

Sellen, D., and R. Mace
1997   Fertility and Mode of Subsistence: A Phylogenetic Analysis. Current
        Anthropology 38:878–889.

1999   A Phylogenetic Analysis of the Relationship between Sub-Adult Mortality
        and Mode of Subsistence. Journal of Biosocial Science 31(1):1–16.

Wood, J. W.
1990   Fertility in Anthropological Populations. Annual Review of Anthropology
        19:211–242.

IV

GOVERNANCE
AND POLICY

The Political Ecology of Dengue in Cuba and the Dominican Republic

Linda M. Whiteford and Beverly Hill

In this chapter, we examine the rapid and accelerating spread of dengue fever in Latin America by applying a political ecology analysis to its control. Dengue, long a mosquito-borne menace in Asia and other parts of the world, was believed to be on the verge of being "controlled" in the Americas. Malaria and yellow fever, two of the best-known examples of a mosquito-borne disease, kill millions each year, while diseases such as dengue fever kill fewer but still weaken and debilitate millions. Although it is impossible to fully gauge the social and economic costs in terms of early and untimely deaths, loss of productive and reproductive potential, and prolonged and debilitating illnesses caused by mosquito-borne disease, its destructive power was clearly recognized as long ago as the early part of the 18th century.

Epidemiology, the study of the determinants and distribution of disease, forms the research basis of public health interventions. The classic epidemiological triangle includes the agent, the host, and the environment (Friedman 1974). The agent may be viral, bacterial, or protozoal, and the host could be a person or animal that is infected. The environment includes sociocultural, political, and geographic variables related to the spread or control of the disease. In the case of dengue fever, the mosquito vector is the mechanism by which the host is infected by the agent. Dengue fever is induced in a human (the host) by means of a puncture bite with which the dengue protozoa (the agent) is transferred from the mosquito vector. While the mosquito does not actually cause the infection, it transmits the protozoa when it takes its blood meal from a human.

According to the traditional epidemiology triangle and public health practice, interventions can be focused toward three primary elements: the host, the vector, or the environment. Focusing on controlling the disease by means of altering the host response to the exposure means the development of a vaccine or other means to immunize populations against the invader. Vaccines against mosquito-borne protozoa have failed due to the extreme complexity of the stages through which the infecting agent moves and its ability to mutate quickly and effectively. The actual agent, in this case the protozoa, mutates quickly, adapting to new hosts, making the control over the actual agent unlikely. Failing to provide a way to immunize the host or control the agent in mosquito-borne disease, public health interventions have moved instead to destroying the vector by controlling the environment.

During World War II, dengue decimated the ranks of the marines invading Saipan, with five hundred men a day becoming sick. The men were often left so weak that they were unable to fight for as long as four and five weeks (Gladwell 2001). Malaria, however, was even worse. According to General Douglas MacArthur, two-thirds of his men in the South Pacific were ill with malaria and 85% of his troops holding Bataan had the disease. In response to the war's demands, a tremendous effort was made to find an effective pesticide. Shortly before the war, a compound known as dichloro-diphenyl-trichloroethane, which came to be known as DDT, had been discovered by a Swiss chemist in the 1930s and seemed to be just the tool for the eradication of the vector. Also in the 1930s, a U.S. public health professional named Fred Soper was working to eradicate malaria by modifying the environment and removing mosquito-breeding places.

Notwithstanding Soper's prolonged and passionate efforts, by 1965 the money from the U.S. Congress to fight malaria ran out, and by 1969 the World Health Organization formally withdrew its support from the malaria-eradication program. The program was expensive, taking much longer than had originally been imagined, and following the 1962 publication of Rachel Carson's book *Silent Spring*, with its indictment of the environmental effects of DDT, its use was curtailed and eventually banned. Without DDT, the campaign to control malaria and other mosquito-borne diseases through the control of its environment stopped. Failure to control either the infectious agent or the vector forced programs to turn to the human hosts to control diseases such as dengue.

Since the 1960s countries have varied in their success at controlling dengue. We suggest that to understand why some countries are more successful than others in reducing the incidence and spread of this disease, we need to use a political ecology approach that expands our analysis beyond the traditional epidemiological triangle to include history and politics. We believe that the human host and his or her physical, cultural, and political environment all need to be considered in order to track the complex constraints that often work against activities to reduce mosquito-borne disease. We apply this holistic approach to a comparative study of dengue in two countries, with our emphasis on Cuba and within the context of globalization. We illustrate that relative isolation from global political–economic trends may, in fact, be associated with successful public health initiatives and positive health outcomes.

THE POLITICAL ECOLOGY OF HEALTH

Human diseases are not static conditions with unicausal etiologies. Whether dealing with chronic or infectious disease, or communicable or noncommunicable disease, the causes are often complex and multifaceted. In today's era of heightened global connectivity, this complexity is even more evident. Medical ecologists have historically not given enough attention to the political and economic aspects of health to suit their counterparts in medical anthropology (Singer 1989). The ecological perspective has been viewed as being too biomedical and less focused on the material aspects of health, although it presents an important aspect in examining outcomes of illness and disease in human populations. The biological aspects of health from an evolutionary perspective have formed the traditional views maintained by many physical medical anthropologists, while medical anthropologists with a cultural focus have made it their goal to identify the social underpinnings of health behaviors and outcomes. Combining the two perspectives gives greater validity to studies that look at human disease, enabling us to take into account the multiple factors at play in the formation of health issues, including processes of globalization.

As other authors in this volume note (Guest and Jones, Armelagos and Harper), political economy is a vital factor related to the health status of a population and can be measured at every level of human organization, from the individual to the society at large. The political economy of health can be summarized in four classic approaches, which include but are not limited to orthodox Marxism, dependency theories, cultural critiques of medicine, and critical phenomenology (see Morgan 1998). Of these approaches, dependency theories are most relevant to this analysis, as they speak directly to the unrelenting inequity manifested as disparity between the developed and underdeveloped areas of the world, which are in turn a product of differential power relationships in the modern global economy. As noted by Lynn Morgan, "dependency theories illustrate the links between what transpires in financial and political capitals and what happens at the farthest reaches of an ever-more-destitute periphery" (1998: 412). Whereas many studies focus on the detrimental effects of globalization on less-developed countries, we go a step beyond and illustrate what one population accomplished because of its relative isolation from global impacts. While we are providing descriptions of political–economic influences on local communities in political–ecological perspective, we point out that, given the right set of circumstances, a community may reject such forces. Similarly, Susan Paulson and colleagues observe that

> while localities are affected by global decision making . . . they are not passive recipients; rather, global environment and development ideas become enmeshed in local struggles in ways that sometimes have larger impacts . . . [on] the application of multiscale research models that bring together selected ecological phenomena, local processes through which actors develop and negotiate environmental management strategies, and global forces and ideas that influence ecological conditions and sociopolitical dynamics. [2003:211]

From a public health perspective, the quote speaks to the benefits of including multiple factors in an analysis, such as evolution, biogeography, ecology, and social science (Levins and Lopez 1999:262). Communities that do not experience the full effects of global forces and are afforded the opportunity of community building with local means are almost perversely benefited by not being dependent on external aid. Instead, the employment of locally organized systems of passive and active surveillance can ensue, which do not perpetuate the cycle of dependency as characterized in theories of political economy.

Dengue control has been approached in multiple ways by researchers and practitioners. The outcomes have proven varied, diverse, and often ineffective.

Community-based programs have failed, along with those established purely via top–down approaches (Kendall et al. 1991). Furthermore, as noted by Duncan Pedersen,

> Environmental constraints are not equally distributed, since political power, access to natural resources and the social production of pollutants is asymmetrically distributed, resulting in different sources of stress and exposure to a degraded environment, thus resulting in different levels of physical, psychosocial and economic harm. This is another reason why the issue of health and disease, human rights and environmental degradation are all inextricably linked. [1996:746]

In the case study we present in this chapter, this theme of the interrelationship between politics, economics, and environmental degradation emerges throughout as the ties between global forces and local health realities become apparent. Our case study demonstrates the theme of dependency in global perspective for one country, while the converse is shown for the other. From an ecological perspective, we look at the problem of dengue and how it is managed with regard to both the macro- and the microlevel forces in two geographically similar, but politically dissimilar, countries.

DENGUE FEVER

Classic dengue fever (DF) consists of four serotypes (DEN-1,2,3,4) and is characterized by fever accompanied by a headache, excruciating pain in the muscles and joints, nausea and vomiting, and rash. It is debilitating but is not as deadly as dengue hemorrhagic fever (DHF) or dengue shock syndrome, both of which may result in death. Dengue is of concern not only because of its debilitating effects on individuals but also because, among the four serotypes, exposure to an earlier infection may render an individual more susceptible when exposed to DHF. DHF, in turn, may cause dengue shock syndrome, from which an individual may die.

The virus is inoculated into humans via a mosquito bite where the mosquito's saliva enters the human body. The virus then replicates itself in the target organs, infects the white blood cells and lymphatic tissues, and is then released into the bloodstream. Transmission occurs when a second mosquito ingests the virus via a blood meal from the infected host. The virus replicates in the mosquito's midgut, and other organs and infects the salivary glands, where it is replicated and passed on to the next human the mosquito bites, forming a cyclical pattern (Centers for Disease Control 2002). To fully understand the existence of dengue as a localized reality, several points must be examined following the epidemiological model of host–agent–vector. First, knowledge of the biological composition of the agent of dengue is a prime element. Second, identification of the place at which the disease vector—the female *Aedes aegypti* mosquito—enters the process of dengue transmission is vital. Finally, and perhaps most important, attention to the articulations between and among the history, political climate, and the interrelationships that manifest at the interface of the human host and their environment is essential.

The *Aedes aegypti* mosquito has adapted to environments that are almost purely human oriented. Unlike mosquitoes that prefer running water, thus predisposing them to rural conditions, the *Aedes aegypti* prefers still water, like that found in flower pots in urban gardens. Therefore, *Aedes aegypti* is much more likely to be found in urban rather than rural areas. To make matters worse, the *Aedes aegypti* is a daytime feeder, more likely to bite while people are out in their yards, on their porches, or gardening. Furthermore, human-made vessels that are typically used for water storage and areas where trash accumulation is common are favorite places of habitation for this insect (Whiteford 1997). Hence, sites where this type of environment exists is a likely place for *Aedes aegypti* to breed, deposit their eggs, and seek out their most desirable nourishment: human blood meal (Adelman et al. 2002).

With between 50 million and 100 million annual cases of DF and DHF worldwide, dengue viruses represent a monumental affliction on developing countries. Latin America and the Caribbean are certainly no exception. While the first outbreak of this infectious disease in the New World was identified in the 15th century in the French West Indies, its prevalence has become a contemporary worldwide pandemic. Dengue epidemics in the Caribbean were reported post–World War II, followed by a reinvasion of the Americas in the 1970s to the 1980s (Gubler and Kano 1997; Whiteford 2000). Subsequent to this incursion, Cuba had the highest incidence of DHF worldwide, after Venezuela and Colombia, respectively, from 1981 to 2001 (Pan American Health Organization [PAHO] 2002). Following a successful campaign to combat a major outbreak of dengue in Cuba in 1981, which included the first major DHF epidemic, outbreaks assaulted this country again in 1997 and 2002.

Dengue, also well known in the Dominican Republic, was part of Soper's early eradication campaign, which continued well into the 1960s. However, as mosquito control programs became increasingly underfunded, they lost their efficacy. Dengue surveillance and control campaigns based on household spraying and inspectors (following the Soper model) dwindled to almost nothing by the 1970s. Community-based surveillance and control campaigns (similar to those practiced in Cuba) also failed in the Dominican Republic for lack of community support. Between 1963 and 1988, the Dominican Republic reported cases of all four serotypes of DF as well as DHF (PAHO 2002). In 2000 there were 3,400 clinically diagnosed cases of DF, 58 cases of DHF, and 6 deaths.

While funding for mosquito control, eradication, and surveillance programs fluctuates, it is important to remember that dengue is not a newly emerging mosquito-borne disease. First identified in China between C.E. 265 and 420, it was associated with flu-like symptoms and often referred to as "bone-break fever" by the joint pain it causes (Whiteford 2000). The first major outbreak in the Caribbean was reported in 1635 in the French West Indies, with other similar cases found in Asia, Africa, and North America by the late 1700s. DHF was recognized in reports by 1780, and by 2000 both DF and DHF existed worldwide (Centers for Disease Control 2000). But the last several decades of the 20th century have experienced an upswing in the number and severity of cases found, particularly in northern South America and in the Caribbean. By 2002 several cases of DHF as well as DF were reported in the United States. Many of the Latin American epidemics appeared after World War II (Gubler and Kano 1997); however, it is difficult to tell how much of that may be attributed to improved diagnostic and reporting systems. What we do know is that the reported number of cases in Latin America accelerated following the war and that the disease showed an invasion of Southeast Asia in the 1950s to the 1960s and a reinvasion of the Americas in the 1970s to the 1980s. The timing of these infections shows a clear association with the cessation of the malaria control programs that depended on the use of DDT. By 1977, both Cuba and Jamaica had significant outbreaks, and the following year both Puerto Rico and Venezuela reported outbreaks.

Between 1981 and 2001, the three countries with the highest reported incidence rates of DHF were Venezuela, Colombia, and Cuba (PAHO 2002). Cases of classic DF are much more difficult to ascertain. DHF is hard to ignore because of the amount of blood loss that is associated with it, often requiring hospital care. Cases of classic DF, however, are much more difficult to assess because the symptoms mimic the flu or other illnesses. Often, people experiencing dengue, while miserable and feeling like they are going to die, do not seek out medical care. Moreover, clinical tests are necessary for proof of infection. In many countries, the reported incidence of dengue is much lower than may actually be the case. In fact, underreporting occurs not only because of difficulties in diagnoses but also because of social and political reasons. Dengue is on the list of diseases that the World Health Organization requires reporting, but many countries underreport levels while some countries fail entirely to report. Countries whose economies are tied to tourism may find it particularly difficult to comply with reporting requirements, especially if the epidemiological system is not well developed. In addition, dengue can be hard to diagnose, and clinical assessments are not common. These are important considerations because they acknowledge that while disease vectors may not recognize political exigencies, those who are responsible for reporting them may.

The lessons we draw from this are (1) a lack of reported cases of dengue does not equal its absence; (2) its long history in the world, if not in the Caribbean, means it is an effective vector; and (3) its control requires a multifaceted, top–down and community–up approach. For these reasons we pay particular attention to the Cuban model, which has resulted in successful reduction in the spread of dengue. We use the Dominican Republic case primarily as a counterpoint.

DENGUE CONTROL AND PREVENTION IN CUBA AND THE DOMINICAN REPUBLIC

To illustrate dengue in Cuba and the Dominican Republic, we turn to review some of the recent outbreaks. Two things stand out: first is the fact that Cuba has been undergoing multiple dengue outbreaks since 1980, and the second is the relative absence of data from the Pan American Health Organization on the Dominican Republic. Remember: The absence of reported cases does not mean the absence of the disease, only a lack of available and reliable data from a given site.

In the 20 years between 1981 and 2001, Venezuela reported 45,799 DHF cases, compared to Colombia's 22,781 cases. Cuba reported 10,586 cases (PAHO 2002). At the same time, Cuba instituted a rigorous environmental surveillance program in its attempt to control DF and DHF outbreaks. It is a program dependent on the unique combination of primary health care, community-based medicine, epidemiology, and neighborhood participation. The political climate in Cuba results in a higher level of compliance than that found in most other countries, which in combination with policies and prac-tices designed to protect the population provides an effective mosquito control program. Compared to the program in Cuba, the mosquito control programs in the Dominican Republic in the 1980s and 1990s suffered from a failure of community participation, a shortage of primary care physicians, an urban-based medical care system that left rural areas with reduced surveillance, and few mosquito reduction field workers (Whiteford 1997, 2000, 2003).

As reported elsewhere in more detail (Coreil et al. 1997; Whiteford 1997, 2000, 2003), global forces such as an economic downturn and the negative economic consequences throughout the Caribbean and Latin America in the 1980s forced the Dominican government to remove or reduce government subsidies on agricultural products and fuels. In response, the powerful trans-portation unions staged "shutdowns" and strikes, culminating in civil unrest and disobedience. During the same period, payment on the external debt be-came a crisis for the Dominican government, forcing it into new discussions and agreements with the World Bank and resulting in debt restructuring and new loans. During the 1980s and 1990s, the Dominican Republic struggled to make payments on loans that were taken out when global institutions made such loans easy to obtain. The consequences of those loans and the recurring debts they incurred forced the Dominican Republic into serious reorganiza-tion of internal policies, reducing the social service sector in favor of an in-creased emphasis on export manufacturing and tourism.

One might ask, what does this all have to do with dengue control policies? The answer is, everything. While there are not data available for dengue incidence in the Dominican Republic that are comparable to those available for Cuba, data are available about prevention programs in the two countries. The political ecol-ogy framework allows us to look at a myriad of factors that, taken together, pro-vide proxy measures to compare the two countries. Global forces, with the Dominican Republic's context being deeply enmeshed in global trade and tariffs, combined with its recent history of democracy and dependency, resulted in the government's move toward the neoliberal economic policies espoused by the in-stitutions of the world banking system. Those policies of privatization and de-centralization resulted in the reduced role of the Dominican government in public health programs of prevention such as dengue control.

As the Dominican government increased adoption of neoliberal economic policies, the number of primary health care workers was reduced; money available for environmental surveillance was cut; and the maintenance of vector-related laboratories fell (Whiteford 1997, 2000, 2003). Centralized services such as sanitation, water, and electricity were in disarray and were the source of constant complaints among the population. In an attempt to increase tourism and prestige, President Joaquín Balaguer widened and repaved some streets and built a powerful lighthouse to attract foreign visitors. However, people complained that the only streets enhanced were those connecting the airport and the lighthouse (El Faro) to the capital and that the renovations displaced many poor and middle-income families who had no other housing. Dominicans were quick to comment on what many saw as a new wave of external control over Dominican policies, this time in the name of globalization.

Research conducted on dengue prevention in the Dominican Republic (Coreil et al. 1997; Whiteford 1997) during this time found that people living in the capital of Santo Domingo had become so disaffiliated from the government that they refused to participate in neighborhood health prevention activities. The research uncovered a vast gulf between what community members believed were their responsibilities in dengue control and the government's obligations. People interviewed in urban neighborhoods said that their government had failed to protect them by failing to provide centralized government services such as reliable sanitation, water, and electricity. The research identified the lack of civil will as implicated in the failure of dengue control activities. Neighborhoods failed to organize around public health themes, and the government neither facilitated nor enforced such organizing. The Cuban example provides a powerful counterpoint to the globalized Dominican context.

Just as the two countries provide striking contrasts, their similarities are undeniable. The Dominican Republic and Cuba both provide similar environments for a range of flowering plants and trees, as well as a host of mosquito vectors. Cubans and Dominicans share a common pleasure in their gardens and patios, often overflowing with scarlet, crimson, and burgundy bougainvillea, brilliant orange acasias, feathery palms, and an abundance of plants in containers. People spend time outside, enjoying their gardens and porches, congregating on the sidewalk to talk or watch, or strolling down the street for ice cream. The common physical environment, along with the many shared cultural practices, makes the comparison between policies and practices adopted by the two countries particularly intriguing. Cuba has experienced several well-documented dengue outbreaks.

In May 1981 Cuba experienced an outbreak with 344,203 cases of DF reported (Khouri et al. 1998), with more than 11,000 cases at the peak of the epidemic. Two-thirds of the deaths (101) occurred in children who were under 15 years old. In all, there were 158 deaths. How was it possible to have a major outbreak of disease and have only 158 people die? What did Cuba do to prevent more deaths? The response of the Cuban government to the initial outbreak was fast, and it occurred early in the process. Because the initial cases were detected relatively early and an existing plan to combat its continued spread was activated, the number of fatalities remained low. Because of early detection, the government moved toward a source-reduction action plan that was heavily dependent on trained human resources as well as the provision of economic resources. Fifteen thousand Cubans were pressed into service to combat the outbreak as inspectors, educators, and container-retrieval and disposal agents. The environmental surveillance program marshaled people as "vector controllers" to get rid of plants that collect water and serve as breeding grounds for *Aedes aegypti*, such as bromeliads and other outdoor plants. Others were trained to handle portable blowers in order to fog homes, and still others enforced sanitary laws concerning the disposal of outdoor water containers. At the same time, the government used airplanes to spray insecticide. This kind of mobilization of resources is possible when policies to combat such an outbreak are developed beforehand, when plans are designed to identify groups to support the effort, and when people are politicized and educated to the public health rationale behind the actions.

According to Gustavo Khouri and colleagues (1998), Cuba's steps to the quick response and effective reduction of potential mortality rates included environmental surveillance, health education, and an open hospitalization policy. The health education campaign used mass media (and a controlled market) to distribute messages about dengue, its signs and symptoms, and its control. The educational campaign built on previously established government-sponsored health education promotions with which the population was already familiar. Perhaps most important and most distinctive from the Dominican Republic's program was the mechanism for engendering community response and support. Since the Cuban revolution, neighborhoods have been organized into politically active units. These neighborhood brigades were thus marshaled in response to the dengue outbreak. Conversely, in the Dominican Republic, attempts to organize communities to participate in cleanup campaigns or to help reduce mosquito breeding grounds were met with either indifference on the part of the residents or distrust of the government's motives (Whiteford 1997). Equally significant, the Cuban government created mobile field hospitals during the crises with a liberal policy of admissions. Again, according to Khouri and colleagues (1998) almost 38% of all the reported cases of dengue were admitted and treated, thus reducing mortality rates significantly. According to official sources and some outside observers, the Cuban response to the 1981 dengue outbreak was innovative and effective (Guzman and Khouri 2002).

There is much to be learned from the Cuban and Dominican experiences. Cuba's political dedication to health is unquestionable. Radical improvements have been made in terms of both health equity and outcome since the revolution. Some of those improvements are in response to the political forces that enforce and reward the creation of a health care system dependent on high numbers of well-trained primary care physicians who remain in neighborhoods. Unlike in the Dominican Republic, where the shortage of primary care physicians has reached critical levels, in Cuba physicians have fewer choices about where and what kind of medicine they practice. Health education in Cuba is a constant, whether on large billboards, over the radio, or in school classrooms. Free public education announcements also occur in the Dominican Republic, but because of commercial media, fewer health education campaigns are aired. Even though the Dominican Republic has state-run low-cost public hospitals, they have become less accessible with the debt restructuring and increased privatization of the 1980s and 1990s.

The 1990s were also economically stressful for Cuba. The Soviet Union was Cuba's major trading partner, and with the former's dissolution Cuba was thrown into an economic crisis. However, because of the Cuban government's continued support for social services, the health care system was not destroyed. While not eliminated during the "special period" of the economic crisis, the level of vector surveillance may have been challenged. Whether surveillance was affected or not, Cuba suffered another dengue outbreak in 1997 with slightly more the two hundred cases of DHF and almost three thousand laboratory-confirmed cases of classic DF (Khouri et al. 1998). There were 12 deaths, but no deaths under the age of 16. It is important to note that between the 1981 and 1997 outbreaks the government established its programs of active and passive surveillance. At the end of the 1981 dengue epidemic, Cuba established a passive surveillance program, including testing suspected patients. No positive cases were identified. Beginning during the 1997 outbreak, the active surveillance program was designed to identify early cases and trace location of transmission. While there is some information to the contrary (Crabb 2001), Cuban data suggest that the early detection prevented the outbreak from spreading to the other 30 municipalities outside its initial site in Cuba. The passive surveillance initiated following the 1981 outbreak showed that mosquitoes were not found in patients' homes and that no indigenous transmission could be established from 1981 to 1996. It also suggested that reinfestation occurred in some areas—for instance, in Santiago de Cuba (where the 1997 outbreak was centered), through the importation of tires transporting *Aedes aegypti* in 1992.

The 1997 outbreak again gave rise to new surveillance techniques. Febrile patients at high risk in the primary health care system in the city of Santiago de Cuba were sought out for further testing between January and July 1997. Sixty thousand emergency room patients were tested between November and January 1997–1998. Home interviews also sought to identify febrile patients at high risk (Khouri et al. 1998). This combination of passive and active surveillance provided information about secondary infections as they related to mortality. Secondary infections were present in 100 of 102 of the DHF and dengue shock syndrome cases. In fatal cases, secondary infections were documented in 11 out of 12 of the cases.

CONCLUSION

The Dominican Republic and Cuba represent two quite distinctive political orientations while sharing a relatively similar ecology. Both are Caribbean islands with histories of indigenous population loss due to colonizing activities of Spain and other European countries; both have current policies and practices influenced by their geographies and political relationships with their large and dominant neighbor, the United States. But the political orientations of the governments are quite distinctive. The Dominican Republic is a democracy based on a capitalist economic system, even though Balaguer, "the last of the Caudillos," was president of the republic throughout many of the 30 years between the downfall of the dictator Rafael Trujillo in 1961 and Balaguer's own death in the 1990s. Cuba also experienced a long period under the dictator Fulgencio Batista, until 1959. Batista was then overthrown by Fidel Castro, who has since maintained power. While the economy of the Dominican Republic is dependent on foreign trade and assistance, much of it from the United States, Cuba's centralized economy was heavily dependent on support from the Soviet Republic, until its dissolution. The U.S. embargo on trade with Cuba was designed to isolate Cuba from global trade. In actuality, it reduces the effect of U.S. practices and policies on Cuba, as they are filtered through global processes. That very reduction may facilitate the Cuba dengue achievements.

A political ecology perspective takes into account how history, culture, politics, populations, and the environment shape policy, practice, and perception. Framing this picture is globalization, that age-old process of trade, transference, and influence peddling, albeit now on a faster, wider, and some say more extensive scale than before. Using that frame, we ask, how can we account for the difference in dengue vector control activities in two countries as similar and dissimilar as the Dominican Republic and Cuba? How does the political ecology perspective enhance our understanding of the rapid and accelerating increased incidence of dengue in the Americas? What have we learned from our analysis and comparison of the recent history of dengue and its control programs in Cuba and the Dominican Republic? We know that official reports of prevalence and incidence rates are not sufficient explanations of what is happening. We also know that even though the vector may not respond to political decisions, the provision of policies and practices, and even the willingness of the population to participate in control strategies, is very much a political process.

In addition, two variables often missing from discussions of vector-borne disease occurrence also contribute to our understanding of the larger picture. They are, first, the relative physical and cultural isolation of the affected population and the degree of its involvement with globalization. The second variable is that of the political commitment to public health infrastructures—both of water and sanitation and of health care. Clearly the two variables are interrelated and, we argue, necessary to any substantive discussion of disease control strategies. Involvement in the global economy redirected resources in the Dominican Republic from the social service sector to the development of income-generating exports and tourism. Cuba, less deeply involved in the global economy and heavily committed to improving health status, continued to support its social service sector, even in times of significant economic crisis. The result is that neoliberal economic reforms initiated by the United States and spread through global financial structures had significant impacts on dengue surveillance and prevention programs in the Dominican Republic. Cuba, on the other hand, suffers economically from its relative distance from the global financial empires dominated by the United States, which, in combination with the Cuban government's sustained commitment to health, results in accelerated rather than reduced dengue surveillance.

The case study presented here shows how a political ecology framework fruitfully extends our analysis of dengue control and prevention by incorporating global processes and the force of globalization while simultaneously maintaining the centrality of history, culture, and the role of the environment. The Cuban model is not flawless, nor is it applicable in many other countries. But by contextualizing its successes we can better conceptualize how prevention campaigns can achieve greater community involvement, a critical element that most public health officials agree is central to vector control. We conclude our analysis with the following quote on dengue:

> Control of dengue transmission is harder today than ever before, but some principles remain fundamental if control is to be achieved: political will (financial support, human resources), improvement of public health infrastructure and vector control programmes, intersectorial co-ordination (partnerships among donors, the public sector, civil society, non-governmental organizations, and the private and commercial sectors), active community participation, and the reinforcement of health legislation. Ministries of health must direct control, and must establish epidemiological and entomological surveillance and education campaigns for the community. It is fundamental that the community recognizes it responsibility in control. [Guzman and Khouri 2002:40]

Times have changed since those of Fred Soper and his eradication teams. The world is more connected now with cell phones and e-mail, and tourism is one of the world's leading industries. But vector control still needs to occur on the ground. Perhaps the clearest lesson learned from the application of the political ecology framework to the question of dengue control is that prevention and control programs require multipronged and multitiered approaches. And successful control programs are still built on the basics that are as true today as they were for Soper over 70 years ago: modification of the environment, reduction of the vector, and early detection in the host.

We should also remember that no community health prevention program succeeds without a politically supported, enforced, and sustained impetus.

The Pan American Health Organization (2002), an organization similar to the World Health Organization but one focused on Latin America, recommends the following five-stage integrated strategy for the control of dengue: (1) develop an integrated epidemiological and entomological surveillance system; (2) encourage advocacy and implementation of intersectoral actions; (3) support effective community participation; (4) generate plans for environmental management; and (5) enhance patient care, inside and outside of the health system. None of these are easy to accomplish, and regardless of what official documents might indicate, most systems fail (Crabb 2001). The barriers to achieving a successful integrated strategy are considerable and include a failure to encourage civil will and community trust in the government, failure to provide secure and reliable water and sanitation systems, failure to develop intersectoral capacity building necessary for coordination and cooperation, and failure to balance the often contrasting needs of individuals with the needs of the public. The task of public health researchers, social scientists, and policymakers is to unite disparate and often competing agendas in order to balance local and national priorities in an increasingly globalized world.

We must also break out of the framework that includes only the biophysical environment in our area of analysis. We should begin viewing attributes that are social, political, economic, and historical in nature, in an effort to avoid oversimplifying the reality of the health problems that we examine. Attending to multiple levels of study, conceptually as well as theoretically, is essential if we are to propose alternatives and solutions that are sound in practice and serve to improve the health status of marginalized populations. Thus, we should continue utilizing the epidemiological model of host, agent, and environment but with a broader view that encompasses the other confounding variables that are at play in the production of environmental health issues.

It seems apparent that Cuba prevailed in meeting the health needs of the immediate community to dengue control during the 1997 outbreak. The health system of Cuba emphasized employing local resources in ways that were ultimately effective. This demonstrates the need for more community-based programs that capitalize on large local mobilization efforts and function with limited monetary resources. This ensures that efforts to improve the health of a community occur within the context of the local history, geography, and politics that therefore have a greater potential for sustainability.

The case of dengue in Latin America is but one example of the political ecology of health in a global perspective. Many others exist, both historically and contemporaneously, and deserve greater attention alongside of the dengue examples. The more attention that is paid to health issues using a political ecology approach, the more we will understand and ultimately ameliorate the costs of needless human morbidity and mortality. This is not to say that there is one right path to understanding human illness and disease but rather to suggest the utility of the myriad approaches, such as those that comprise the political ecology approach.

REFERENCES

Adelman, Zachary N., Nijole Jasinskiene, and Anthony A. James
2002   Development and Applications of Transgenesis in the Yellow Fever
        Mosquito, Aedes aegypti. Molecular and Biochemical Parasitology
        121:1–10.

Centers for Disease Control
2002   Dengue Fever. Electronic document, www.cdc.gov/ncidod/dvbid/dengue,
        accessed on June 30, 2002.

Coreil, Jeannine, Linda Whiteford, and Diego Salazar
1997   The Household Ecology of Disease Transmission: Dengue Fever in the
        Dominican Republic. In An Anthropology of Infectious Disease:
        International Health Perspectives. Peter Brown and Marcia Inhorn, eds.
        Pp. 145–171. Amsterdam: Gordon and Breach.

Crabb, Mary Katherine
2001   Socialism, Health and Medicine in Cuba: A Critical Re-Appraisal. Ann
        Arbor: University of Michigan Dissertation Services.

Friedman, Gary D.
1974   Primer of Epidemiology. New York: McGraw-Hill.

Gladwell, Malcolm
2001   Annals of Public Health: The Mosquito Killer. Reprinted from the New
        Yorker. Electronic document, www.gladwell.com/2001/2001_07_02_a_ddt
        .htm, accessed March 27, 2005.

Gubler, D., and Kano, G.
1997   Dengue and Dengue Hemmorhagic Fever. New York: CAB International.

Guzman, Maria G., and Gustavo Khouri
2002   Dengue Update. Lancet 2:33–42.

Kendall, Carl, Patricia Hudelson, Elli Leontsini, Peter Winch, and Linda Lloyd
1991   Urbanization, Dengue, and the Health Transition: Anthropological
         Contributions to International Health. Medical Anthropology Quarterly,
         New Series 5(3):257–268.

Khouri, Gustavo, Maria Guadalupe Guzman, Luis Valdes, Isabel Carbonel, Delfina
   del Rosario, Susana Vazquez, Jose Laferte, Jorge Delgado, and Maria V. Cabrera
1998   Reemergence of Dengue in Cuba: A 1997 Epidemic in Santiago de Cuba.
         Emerging Infectious Diseases (4)1:89–92.

Levins, Richard, and Cynthia Lopez
1999   Toward an Ecosocial View of Health. International Journal of Health
         Services 29(2):261–293.

Morgan, Lynn
1998   Latin American Social Medicine and the Politics of Theory. In Building a
         New Biocultural Synthesis: Political-Economic Perspectives on Human
         Biology. Alan H. Goodman and Thomas L. Leatherman, eds. Pp. 407–424.
         Ann Arbor: University of Michigan Press.

Pan American Health Organization
2002   Dengue Fever in the Americas: Update. Electronic document, www
         .hc-sc.gc.ca/pphb-dgspsp/tmp-pmv/2002/df0327_e.html, accessed
         on June 30, 2002.

Paulson, Susan, Lisa L. Gezon, and Michael Watts
2003   Locating the Political in Political Ecology: An Introduction. Human
         Organization 62(3):205–217.

Pedersen, Duncan
1996   Disease Ecology at the Crossroads: Man-Made Environments, Human
         Rights and Perpetual Development Utopias. Social Science and Medicine
         43(5):745–758.

Singer, Merrill
1989   The Limitations of Medical Ecology: The Concept of Adaptation in the
         Context of Social Stratification and Social Transformation. Medical
         Anthropology 10:223–234.

Whiteford, Linda

1997   The Ethnoecology of Dengue Fever. Medical Anthropology Quarterly
         11(2):202–223.
2000   Local Identity, Globalization and Health in Cuba and the Dominican
         Republic. *In* Global Health Policy, Local Realities: The Fallacy of the Level
         Playing Field. Linda Whiteford and Lenore Manderson, eds. Pp. 57–78.
         Boulder, CO: Lynn Rienner.
2003   Dengue Fever: The Success of Surveillance. Presentation at the Centers for
         Disease Control and Prevention, Atlanta, July 29.

International Architecture for Sustainable Development and Global Health

Paul R. Epstein and Greg Guest

International initiatives to reformulate world order have often followed cataclysmic events, such as world wars, widespread social unrest, pandemics, or worldwide depression. Today the destabilization of global environmental systems and the accompanying global health crisis are driving us toward another interval of reflection and the reordering of priorities and policies to reprogram international development.

When representatives of multiple nations have jointly established new agreements, rules of engagement, institutions and incentives, many such "intervals" have often resulted in renewed international hierarchies, sowing the seeds for subsequent conflicts. The 1884 Berlin Conference (Hochschild 1998), at which colonial powers divided Africa into separate dominions, created antagonisms that contributed to the First World War. In 1919, following the war, the Paris Peace Conference reshuffled colonies, designating them as "mandates" under the League of Nations and setting the stage for discordant development, depression, and the Second World War (Macmillan 2003).

Following two world wars, the 1929 crash of the stock market, and the Great Depression, the Bretton Woods Conference was held in 1944 against the backdrop of the White Mountains of New Hampshire. It established a new world order that ushered in a period of prosperity and the subsequent independence of most of the old colonies. But the unraveling of the Bretton Woods rules in the early 1970s sparked a spiral of inflation, insurmountable debts and speculation, and widening divisions between wealthy and poor nations.

Today climate instability and the exhaustion of resources (forests, soils, water, biodiversity), together with growing inequity and deepening poverty, are resulting in the emergence, resurgence, and redistribution of infectious disease, stalking humans, plants, and animals (Epstein et al. 2003). The conditions are not sustainable, and the mounting social and economic costs are creating convergent agendas among members of civil society, international institutions, and the economic sector. New rules and incentives, along with reprogrammed institutions are needed to create a more equitable, healthy, clean, and sustainable form of development.

In this chapter, we suggest an alternative framework for sustainable development and subsequent global health. We begin with a brief review of some of the ecological exigencies with which our planet is faced, focusing on global climate change and the relationship these conditions have to human health. We then provide a brief history of "world orders" as a backdrop to the current situation in which we find ourselves, and we conclude with a proposed outline for a new global financial architecture.

GLOBAL ECOSYSTEM CHANGE AND HUMAN HEALTH

As Greg Guest and Eric C. Jones summarize in the introduction to this book, natural resources are being exploited at an alarming and unsustainable rate. Simultaneously, chemical waste products of energy production and manufacturing are altering the heat budget among the atmosphere, land surface, oceans, and ice cover, altering the hydrological cycle and weather patterns. We inhabit an increasingly globalized society, where events and actions in one part of the world can no longer be viewed in isolation; we share the air we breathe, the oceans and soils from which we derive nourishment, and the earth's climate system. Yet, our current use of these "global commons" does not reflect our long-term needs (Dietz et al. 2003).

Changes in atmospheric chemistry and planetary physics are affecting biological and social systems, with the combined changes having direct and indirect impacts on human health and well-being (Patz et al. 1996; Rosenzweig et al. 2001; Watson and McMichael 2001). Spreading disease and malnutrition have social and environmental determinants that affect water quality and availability, agricultural productivity, and the condition of forests. The same set of conditions is leading to infections in wildlife, livestock, agricultural systems, forests, and coastal marine ecosystems (e.g., coral reefs; Epstein et al. 2003). The potential for rapid dissemination of infectious diseases incubated anywhere on the globe—HIV/AIDS, severe acute respiratory syndrome (SARS), influenza, bovine spongiform encephalopathy (mad cow disease), West Nile virus—and their ability to disrupt trade, travel, and tourism provide impetus to address the conditions that underlie ill health everywhere.

Concomitant with this ongoing environmental degradation is an increasing gap in wealth around the globe, both within nations (Di Carlo et al. 1997; Victora et al. 2000; Lynch et al. 1997) and among them (United Nations Development Programme 1999). Today one-quarter of the world's population lives in extreme poverty (United Nations Development Programme 1997); it is also the poor who are most vulnerable to the detrimental effects of environmental degradation and climate change (Houghton et al. 2001). Having neither the ability to prepare for, nor the adequate resources to cope with, the consequences of global environmental change is profoundly affecting the quality of life for many of the world's citizens.

Poverty is a major predictor of overall poor health status. Today, as many as one-fifth of the world's citizens (about 1.3 billion people) lack access to safe drinking water, and about one-third lack access to proper sanitation (Tong et al. 2002). The absence of these basic needs alone causes millions of preventable deaths worldwide each year (Esrey et al. 1991). Malaria, tuberculosis, and HIV/AIDS kill over six million people annually, the majority of whom live in developing areas, where funds and health care systems are often inadequate. These diseases also add to population vulnerability and diminish local productivity (World Health Organization 2002).

The World Health Organization (1997) estimates that poor environmental quality is directly responsible for about 25% of the world's burden of disease and that marginalized populations are most vulnerable to these effects. Poverty and degraded ecological systems provide ideal breeding sites for the emergence of infectious diseases and their vectors, such as insects, rodents, and bats. With habitat loss and fragmentation, opportunistic vector populations can explode, particularly following sequences of weather extremes that reduce predators and boost prey. Stressful conditions, malnutrition, and great

burden of disease may even encourage the evolution of pathogens from be-
nign to malevolent organisms through accelerated mutation rates and de-
pressed immune surveillance systems (Beck 1995; Bjedov et al. 2003).

Not only do those populations with the fewest resources have to cope with
unrelenting vector- and water-borne disease, but they also bear the majority
of direct health risks associated with intensified weather extremes increasingly
linked to human activity—for example, heat waves, prolonged droughts, and
floods (Epstein 1999; Stephens et al. 2000). In August, the World Meteorological Organization (2003) reported that increased heat in the global climate
system accounted for the severe and erratic weather patterns being observed
throughout the globe. In December 2003, the World Health Organization estimated that 160 thousand annual deaths are now attributable to climate
change and are projected to increase in coming years and decades.

Climate Change and Severe Weather

Weather is largely driven by ocean conditions. In the past 50 years, the temperature of the ocean water to a depth of two miles (i.e., halfway to the
seafloor) has increased (Levitus et al. 2000). Warming of the atmosphere is occurring disproportionately at high latitudes (Houghton et al. 2001), and the
combination is changing the North Atlantic (Hurrell et al. 2001; Hoerling et
al. 2001). In summers since the 1970s, North Polar ice has shrunk from ten
feet to five feet in thickness, while Greenland has lost 9% per decade. The
North Atlantic Deep Water Pump—the cold *salty* sinking water that pulls the
Gulf Stream north—is slowing, as thawing ice and more rainfall at high latitudes leave cold *fresh* water that does not sink. At the same time, the tropical
Atlantic is becoming saltier, as ocean warming increases net evaporation and
rainfall at high latitudes (Curry et al. 2003). North Atlantic pressure gradients
versus warmed subtropical seas are affecting weather in the northeastern
United States and Europe. Ice-core records indicate that "cold reversals" have
interrupted warming trends in the past (National Research Council 2002),
and that abrupt change is the norm, not the exception.

No nation is immune to the impacts of climate change. Extreme weather
events such as thermal extremes, droughts (and wildfires), and severe winter
conditions are increasing in frequency in North America and Europe (Climate
Change Futures, 2005), as are the associated economic costs. The causes of environmental distress are also global in scope, although developed nations share

the majority of responsibility. With no letup in the burning of fossil fuels and in the introduction of other greenhouse gases, we are forcing the planet's climate into uncharted waters (Levitus et al. 2000; Barnett et al. 2001). In fact, recent estimates indicate that the earth could soon become warmer than at any period in the past 1 million to 40 million years (Houghton et al. 2001).

Climate instability and the accompanying threats to life-support systems on which human health and well-being depend require a coordinated global response. Despite being subsumed by transnational corporations, concerns about loss of sovereignty must be suspended in order to protect the basic environmental elements and life-support systems—temperature, pressure gradients, winds, the hydrological cycle, large-scale ocean circulation systems, coral reefs, and the globe's ice cover—all of which are currently undergoing rapid and potentially nonlinear changes.

Current forms of international governance are inadequate to address the tragedy and potential collapse of the commons. The increase in global health threats has spurred international institutions—UN agencies and trade organizations—to develop funds to treat the diseases. In addition, several international treaties address some of the root causes. But while enabling funds were crucial for nations to accept the 1987 Montreal Protocol to control stratospheric ozone-depleting chemicals, the Kyoto Protocol—aimed at controlling greenhouse gases and growing forests as carbon sinks (i.e., land area with vegetation that absorbs carbon dioxide)—does not have the magnitude of financial incentives that it needs for universal acceptance. Establishing a new international economic order will require collaboration among society, UN agencies, scientists, and the economic sector to derive the principles, programs, practices, and policies that will drive the clean energy transition.

A Brief History of World Orders

Conscious international efforts to reorder rules of engagement have often followed periods of social turbulence, pandemics, revolutions, depressions, and wars. The following is a brief survey of some such initiatives taken over the past two centuries.

In the late 18th century, revolutions against the *anciens régimes* exploded in the Americas and in France (1789), with declarations of *The Rights of Man*. Napoleon, after his defeat in Haiti (which led to his capitulation of American territory with the Louisiana Purchase in 1803), marched across Europe. Following his final defeat, the 1815 Treaty of Vienna was signed to reestablish order in Europe; but there was no representation from nations outside of Europe, and international relations continued to be dominated by colonial pursuits.

By mid–19th century, urban crowding (the population of London rose sevenfold from 1790 to 1850) overwhelmed water and sanitation systems.[1] In the 1830s outbreaks of cholera, tuberculosis, and smallpox in London, New York, and Boston (and many other industrializing cities) affected city dwellers and sparked protests and revolutionary movements demanding change. Public health jumped to center stage in national development agendas during this period, and the resulting sanitary and environmental reforms had stemmed the tide of the infections by the 1870s, years before Robert Koch (1883) and Louis Pasteur (1890s) isolated *Vibrio comma*, *Mycobacterium tuberculosis*, and *Bacillus anthracis*. While no concerted international developmental effort emerged during this period, the first international funds to coordinate communications were established to support the International Telegraph Union (1865) and the Universal Postal System (1874).

The next turning point came in 1884, when colonial powers met in Berlin to consolidate a new world order. After centuries of extracting gold, ivory, and slaves, colonial powers divided Africa into new dominions (Hochschild 1998). During the succeeding decades, primary accumulation (from colonies) and concentration of wealth were accompanied by large migrations of peoples, populist and labor movements, a failed revolution in 1905, the First World War, the Russian Revolution of 1917, and the pandemic of influenza (killing some 20 million to 40 million). It was time to reorder international relations once again.

The victorious powers from World War I—led by the United States, the United Kingdom, and France (Woodrow Wilson, David Lloyd George, and Georges Clemenceau, respectively)—met in Paris for a peace conference in 1919, which would last from January to June (MacMillan 2003). The Treaty of Versailles and the League of Nations (following Wilson's "Fourteen Points" and principles of "self-determination") kindled new hopes for a new order and a lasting peace. But there were many problems left unsolved. Women and women's suffrage were rejected; reparations were exacted from the vanquished; and colonies were reshuffled into League of Nation "mandates." Without funds to stimulate growth, it took six years for Europe to recover, and the seeds of future conflict smoldered. Rapid, uncoordinated growth led to spirals of speculation. Following the 1929 stock market crash, the Bank for International Settlements (established in 1930) served as a hint of what was to come following the Great Depression and yet another world war, more devastating than the first.

In 1944, beleaguered, war-weary world leaders convened in Bretton Woods, New Hampshire. There, determined to avoid the pitfalls of the past, they crafted a new framework for international relations and development. It is essential to note that vast areas of the globe—Asia, Africa, and Latin America—were invisible in these deliberations, omissions that would later hinder sustainable solutions.

Under the guidance of John Maynard Keynes, the British lord who had secured U.S. financial support for the war in Europe through the "Lend–Lease Act," the Bretton Woods agreements initiated a new era by resetting the monetary signals that redirected markets, financing, and private enterprise. Three rules of trade were established: (1) free trade in goods but (2) constraints over the trade and flow of capital investments and (3) fixed currency exchange rates.

New institutions were established to stabilize the post–World War II order and stimulate recovery and growth. The International Bank for Reconstruction and Development (the World Bank) was established in 1944 to leverage long-term investments, while its sister organization, the International Monetary Fund (IMF), was established in 1945 to provide short-term stabilizing funds and policies. The United Nations was launched soon after, followed by the United Nations Children Fund (UNICEF) in 1946. Under the guidance of Eleanor Roosevelt, the Universal Declaration of Human Rights framed the moral structure for the new world order.

But recovery would have been difficult and slow without new financial incentives. In 1947, the European Recovery Program, or "Marshall Plan," gave Europe the support it needed to revive European economies and thus have its countries resume their positions as healthy trading partners.

Within the United States, the GI Bill provided an enormous boost to the national economy, subsidizing housing, education, and jobs. Other subsidies followed, notably for airports and highways, as well as coal and oil exploration, thereby catapulting air and ground transport as primary drivers of the postwar economy.

In 1972, as the Vietnam War stretched resources and as military spending drove up deficits, U.S. president Richard Nixon came under increasing political pressure. It was then that the Bretton Woods agreements were abandoned. During this period the price of gold jumped from $35 an ounce to over $600, and with the help of the "oil crisis," the price of a barrel of oil rose from $3 to $30. Currencies were allowed to float, boosting the speculative betting on their relative values.

As the price of imported oil skyrocketed, developing nations became severely strapped by their dependence on oil to run their transport systems and economies. They went to the World Bank and IMF for loans. The Bank had plenty to loan, as the "petrodollars" flowed in and many nations borrowed and invested in large-scale development projects such as dams. *Import substitution* became the buzzword, as developing nations were encouraged to industrialize. Unfortunately, aside from the four "Asian Tigers" (Korea, Hong Kong, Singapore, and Taiwan), industrialization did not occur. And monoculture agricultural production designed for export (to repay the loans) remained the primary economic driver, bringing with it exhaustion of soils, deforestation, and loss of food security, as subsistence-based farmers replaced traditional and life-sustaining farming practices with less-predictable cash cropping.

In the 1970s a group of developing nations called for a new international economic order (NIECO). The NIECO would include debt cancellation and restructuring of international terms of trade (more on this later). But Western nations rebuffed this initiative.

By 1983, the money flowing out of the developing world surpassed that flowing in, and high interest rates and hyperinflation (especially in Latin America) deepened the debt crisis during the rest of the decade.

The IMF instituted structural adjustment programs during this period, designed to boost export crops in developing countries and to reduce state spending. But the belt-tightening stipulations ("conditionalities") that cut nation-state spending on housing, education, health, and public transport emasculated the very sectors that buttress domestic economies and provide a modicum of social support. Dismantling state infrastructures has harmed many nations and has widened income gaps. The discrepancy between the "state of the economy"—the gross national product of nations—and the condition of vast populations has continued to grow.

By the early 1990s it became apparent—even to prominent World Bank economists—that structural adjustment programs were undermining national infrastructures, preventing poverty alleviation and competition, as well as encouraging corruption. Something had to be done to re-create the social nets and prevent more "failed" states. But little was done (save the issuing of empty vouchers for food and other basic needs).

Meanwhile, floating exchange rates facilitated an explosion in currency speculation, and trading rose from $18 billion daily in 1972 to over $1.5 trillion a day in the 1990s! The rapid movement of what might be called "specu-dollars" helped set the stage for the 1996–1997 collapse of the Bhatt in Thailand and other Asian currencies. (Malaysia was spared by restricting capital flows. Roundly criticized by the World Bank and IMF at the time, their foresight later gained praise.) The collapse ushered in an intense four-year period of speculation, the Enron debacle, and deceit by Wall Street analysts who knew that the material basis for many stocks had evaporated but still encouraged the average investor to buy.

But most economists held fast to the belief in the power of the unregulated market to right all ships ("market fundamentalism"). The "Washington Consensus"—the drive toward liberalization of trade, deregulation, and, privatization—continued as the dominant philosophy, with no distinction made between the trade in goods and the movement of capital, technologies, and people.

The search for cheap labor and cheap goods has driven trade and transactions in the international economy for over five hundred years. For the producers of raw materials, the former has meant consistently low wages and questionable benefits for many nations rich in resources. The rapid flow of capital—moving in and out of nations in minutes to days, rather than serving as long-term investments—precipitated the bursting of "the bubble" in the late 1990s. Low prices for raw goods and basic commodities—maize, coffee, and nuts—have persisted, while prices for imported goods and technologies have risen sharply. Profit margins associated with globalized commerce have become evermore concentrated in multinational corporations that move all elements of trade with minimal regulation and—until 2003—with maximum protection by the World Trade Organization (WTO).

The WTO is an international organization dealing with the rules of trade among nations, and it was established in 1995 to replace the General Agreement on Tariffs and Trade (GATT), extending trade policy from goods to service industries through the General Agreement on Trade in Services (GATS). The WTO was formed to ensure that international trade "flows as smoothly, predictably and freely as possible" and to "help producers of goods and services, exporters, and importers conduct their business" (WTO 2004). Unlike the GATT, the WTO is backed by a dispute settlement mechanism, or trade court, the findings of which are mandatory for member governments. Through this court, the WTO acts as an arbiter when members' nontrade objectives conflict with their free-trade undertakings. The effectiveness of the WTO, however, has been questioned, as has the impact of its policies on those most in need.

WTO talks have broken down several times in the past four years. In 1999, the talks in Seattle collapsed from internal irreconcilable divisions and the weight of protests directed at labor and human rights, the environment, and secrecy. The talks in Cancun, Mexico, in 2002 saw a similar fate, due to sharp differences of opinion between wealthy and poor nations.

The record of the WTO with respect to global health does not fare much better. In the health arena, the British journal *Lancet* describes how deregulation and liberalization have meant siphoning national budgets in developing countries to establish foreign-owned health maintenance organizations (Pollock and Price 2000) and preventing nations (e.g., Thailand) from producing inexpensive HIV medicines (Loff 2002). The former is reprehensible; the latter is unconscionable. The WTO has also played a negative role with respect to making HIV medications widely available, protecting the patent rights of international pharmaceutical companies under the Agreement on Trade-Related Aspects of Intellectual Property Rights (TRIPs), which governs the content of domestic patent law and how it influences access to pharmaceuticals. In this regard the WTO has operated with little foresight as to the long-term social and economic impacts of HIV and implications for international trade, as millions face disease, loss of key societal leaders, and declining productivity (Loff 2002). But, in late 2003, the WTO began to lift restrictions on development of essential medications, changes that have affected Thailand, India, South Africa, and Brazil.

Macroeconomic forces in general have played a major role in retarding health advances. For instance, there is systemically entrenched unemployment of essential workers; 4,000 nurses, 100 clinical officers, 2,000 laboratory staff, and 160 trained in pharmacy are jobless in one country alone, Kenya (Lee 2003). The structural adjustment programs and other "conditionalities" to maintain credit lines have meant that salaries for trained personnel are totally inadequate, and this continues to drive the "brain drain" from developing nations. In addition, it greatly diminishes the efficacy of new moneys and programs and further taxes the already stressed and underfunded medical systems in these countries.

AN ALTERNATIVE ARCHITECTURE

While grievances among nations during WTO talks in Seattle coalesced into new coalitions, an extensive structure of underlying roots of the present order was unearthed: the perverse incentives favoring transnational corporations and the vast inequality in the prices of goods traded (unequal "terms of trade") reinforce the overwhelming economic power of developed nations (Epstein 2000).

Moreover, with the restructuring of world power relationships following the collapse of the Soviet Union, the interests of the United States have ascended, as Europe seeks to form an alternative pole. And within the United States in the new century, the interests of the oil industry have become dominant, reinforcing its role as a major source of large-scale environmental damage and the greenhouse gases altering the climate. To the detriment of many other economic interests in the financial sector (insurance, banks, institutional investors, and accounting firms) and in the industrial and farming sectors, the dominance of oil has created a crisis within the U.S. economy.

Creating a system that promotes healthy economic development will require creating new "rules" and new funds and transforming institutions. The current international economic "system" does not support the sustainable use of global common resources—fisheries, oceans, fresh water, and atmosphere. In addition to promoting oil interests, the current regulatory framework continues to favor transnational corporations and current trade agreements, at the expense of environmental stability and global public health (Stephens et al. 2000).

Some progress, at least on paper and in spirit, has been made. Based on the Millennium Summit in 2000, the United Nations drafted a declaration outlining eight "Millennium Development Goals" related to improving global health, reducing poverty, and achieving development that is equitable and environmentally sustainable (United Nations 2002a). Attendees of the 2002 Summit of the Americas in Monterrey, Mexico, and the 2002 World Summit on Sustainable Development in Johannesburg (United Nations 2002b) signed declarations that promoted fundamental principles similar to the Millennium Development Goals. In spite of this progress, members disagreed on specifically how to achieve such goals. The result was pronouncements that lack enforceable guidelines and adequate financial incentives.

Advances have been made on the public health front. Recently created philanthropic organizations, such as the Bill and Melinda Gates Foundation and the Global Fund, have made significant financial commitments toward improving the health of citizens in developing countries and represent noble efforts to mitigate the burden of malaria, tuberculosis, and HIV/AIDS in poorer regions of the world.

Grassroots organizations have also made some progress in bringing health problems of the poor to the global stage and to the attention of policymakers (Werner 1998). Nonprofit organizations such as the Global Health Council have been successful in mobilizing thousands through grassroots efforts and in linking local health realities with international institutions. Nowhere is this more apparent than with HIV/AIDS activism, which is vociferous and encompasses multiple dimensions of the disease (e.g, stigma, access to care, education, research ethics). Unprecedented global and multilateral discussion of HIV/AIDS issues was facilitated at the 2004 International AIDS Conference in Bangkok. Community meetings, exhibitions, and forums in the conference's "Global Village" provided a bridge between conference participants and community-based groups, enabling them to share experiences and ideas.

While well intentioned and necessary, philanthropic and grassroots efforts need to be supplemented with global initiatives and policies that provide developing nations with the ability to fund their own programs and bolster public health on their own terms. There are glaring gaps in health care system infrastructures, specific training, and rational use of health care personnel for the prevention and care of HIV/AIDS. Conditions placed on aid by the Bretton Woods institutions—the World Bank and IMF—undermine many of the aid efforts, by cutting state spending on health, housing, and education (Stephens et al. 2000). A new, specific, and actionable architecture is needed to reduce the disparity in wealth between nations and ensure that economic development does not come at the expense of the global ecosystem and the well-being of marginalized populations. Such an architecture for global governance could have three basic elements: new trade rules, a new administrative framework, and incentives and funds.

Rules and Regulation

Financial reprogramming is needed to drive sustainable development. New rules are needed to (1) raise wages throughout the world and provide fair prices for basic goods; (2) constrain capital flows to prevent the volatile, destabilizing, speculative movement of capital; and (3) direct funds toward healthy development. Furthermore, unpayable and unrecoverable debts must be forgiven. Debt forgiveness would be compensation for past inequities of trade terms and wealth extraction. Projects aimed at forest preservation, for example, cannot keep pace with debt-driven timber extraction, land clearing for monocultures, and the desperate need for fuel. But the debt will rapidly reaccumulate unless the conditions that gave rise to it are themselves changed. Last, interest rates need to be optimized. Rates of 4% or 5% may be too low for finance; rates in the teens are clearly harmful for the productive sector and the majority of the population. Rates of 6% or 8% might be the range to provide the optimum incentive for the two major economic sectors.

The most challenging issue is the terms of trade: the differences between the prices poor nations receive for their exports (e.g., food) versus the prices they pay for imports (e.g., tractors). Since the 1960s, terms of trade have steadily widened. Equalizing such terms is most challenging because it means distributing global wealth more equitably. Terms of trade must become more balanced, even from a Western, self-enlightened perspective, in order for nations to maintain purchasing power in global markets. Equity is prerequisite for trade that is free and fair and for addressing international divisions among workers.

Institutions

The World Bank is a bank and must generate a return on investment (profit). It is therefore not, as presently constituted, the most suitable institution to initiate policy aimed at the public good. The World Bank and the World Trade Organization would have to drastically alter the interests they represent and reformulate their activities to complement broader, democratically developed objectives, such as the Millennium Development Goals.

One possible candidate for administrating global governance is the Global Environmental Facility (GEF)—jointly supported by the United Nations Development Programme, the United Nations Environmental Programme, and the World Bank. The Global Environmental Facility gives grants, not loans. Recently it has increased nongovernmental organization participation. This dimension is still inadequately represented, and the funding of the GEF is grossly insufficient.

Incentives and Funds

Perverse subsidies—those encouraging the extraction, mining, refining, and combustion of coal and oil—must be eliminated. Subsidies and tax incentives must be switched to stimulate producers and consumers of clean energy and energy-efficient technologies. New enterprises for fuel cells, solar panels, wind mills, green buildings, and smart growth can generate jobs and trade—a "win–win" for the economy and the environment.

International agreements—such as the Kyoto Climate and Biodiversity Conventions—are hampered by the paucity of financial resources. Universal acceptance of the 1987 Montreal Protocol to phase out stratospheric ozone-depleting chemicals was achieved when funds were allocated to transfer technology to poor nations. For wealthy and poor nations, funds can help jump-start clean, infant industries. Such agreements must also be supported and honored by those nations that are the biggest producers of greenhouse gases.

Funds are also needed to support what the private sector will not, such as watershed protection and sustainable extraction of resources. Compensation will be needed for nations sharing their genetic resources for medicines and crops. And funds are needed to develop vaccines and drugs for diseases such as HIV/AIDS and malaria, for which there is no current lucrative market in the most afflicted nations (World Health Organization 2001). Money is also needed for reparations for climate and extreme weather-driven devastation in nations such as Honduras, Venezuela, and Mozambique. Extreme weather events are having long-lasting ecological and economic impacts on a growing cohort of nations, affecting infrastructure, trade, travel, and tourism.

An international fund is needed to effectively deal with global environmental and social issues. Potential sources for such a world development fund include taxes on carbon, airline traffic, and Internet "cyberdollars." A tax on currency transactions—the Tobin tax (after the Nobel–prize winning Yale economist James Tobin)—can deliver a "two-for": it could help stabilize finances *and* generate significant funds. A quarter of a penny levy on each of the $1.5 trillion traded daily—far less than one pays a broker to buy stocks—would not discourage long-term investments and would yield over $100 billion annually!

Finance (insurance, banks, institutional investors) that has the longest time-line on the trajectory of costs projected with increasing climate instability can begin to reset credit and insurance signals to help steer industry. Ultimately, when global events compel the global community to change course rapidly, incentives (e.g., altered tax structures, subsidies, and government practices) must be harnessed to change how, not whether, we develop.

MOVING FORWARD

Groups converged in Seattle at the close of the 20th century, sharing multiple grievances against the "corporate agenda" (Anderson and Cavanaugh 2000). Remarkably, 1,448 nongovernmental organizations signed a common statement on equity, poverty, and the environmental impacts of the nonsustainable patterns of production and consumption. Alongside U.S. labor's understandable focus on American jobs, the closing of West Coast ports by the International Longshoremen's Association in sympathy with the demonstrators sent shivers across Washington. Now local environmental initiatives—involving mayors and governors—are springing up to tackle these issues. A force consisting of nongovernmental organizations and citizen groups has emerged, and many businesses are joining these forces.

Worldviews can shift rapidly (Gladwell 2002). At the World Economic Forum held in Davos, Switzerland, in January 2000, the CEOs from over one thousand corporations and the heads of most of the world's governments voted overwhelmingly that climate change posed the greatest threat to all nations. Ideally, public acknowledgment of other impending environmental disasters will follow; even better, large corporations and world leaders may soon begin to consider equity and ecological health as prime directives in the development equation.

Currently, however, concepts such as environmental health, equity, and public well-being are considered "externalities" in neoclassical economics. At best, such principles take second place to the bottom line (i.e., profit margins) and investor satisfaction (which is typically a function of the former). Discount rates may apply to commodities, such as cars and refrigerators, but the value of ecosystems, Earth's resources, and life-support systems rises. Over the past decade a small group of ecological economists has strived to correct this short-term perspective (e.g., Costanza 1991; Daly 2000; Kohn 2004), but such models must be adopted outside of academia and used to guide corporate, government, and international policies.

Constructing a new international economic order with a resilient regulatory, institutional, and financial framework will require a "Bretton Woods II" summit—this time with representatives from the economic sector, civil society, scientists, the United Nations, and some government representatives. A new architecture for governance can constitute the scaffolding to sustain healthy, ecologically sound, and equitable global systems. The clean energy transition can be the engine of growth for the 21st century, propelling us into a healthier, more equitable, and sustainable future. Greater equity and healthier ecosystems can lead to positive health outcomes for all.

NOTE

1. Interestingly, in C.E. 541, similar conditions accompanied the collapse of the Holy Roman Empire, and plague hit European cities, contributing to their abandonment and eventually to feudalism. In C.E. 1346, plague reappeared and rampaged through filthy European cities, this time propelling protests and vast changes in business, science, and technologies (Epstein 1992).

REFERENCES

Anderson, Sarah, and John Cavanagh
2000   Top 200: The Rise of Corporate Global Power. Washington, DC: Institute for Policy Studies.

Barnett, Tim P., David W. Pierce, and Reiner Schnur
2001   Detection of Anthropogenic Climate Change in the World's Oceans. Science 292:270–274.

Beck, M. A., Q. Shi, V. C. Morris, and O. A. Levander
1995   Rapid Genomic Evolution of a Non-Virulent Coxsackievirus B3 in Selenium-Deficient Mice Results in Selection of Identical Virulent Isolates. Nature Medicine 1:433–436.

Bjedov, Ivana, Olivier Tenaillon, Benedicte Gerard, Valeria Souza, Erick Denamur, Miroslav Radman, Francois Taddei, and Ivan Matic
2003   Stress-Induced Mutagenesis in Bacteria. Science 300:1404–1409.

Climate Change Futures
2005   Health, Ecological and Economic Dimensions. Project of the Center for Health and the Global Environment, Harvard Medical School, the Swiss Re-Insurance Company, and the United Nations Development Programme.

Costanza, Robert, ed.
1991   Ecological Economics: The Science and Management of Sustainability. New
        York: Columbia University Press.

Curry, R., R. Dickson, and I. Yashayaev
2003   A Change in the Freshwater Balance of the Atlantic Ocean over the Past
        Four Decades. Nature 426:826–829.

Daly, Herman
2000   Ecological Economics and the Ecology of Economics: Essays in Criticism.
        London: Edward Elgar.

Di Carlo, A., M. Baldereschi, and L. Amaducci
1997   Socioeconomic Inequalities in Morbidity and Mortality in Western Europe.
        Lancet 350(9076):518.

Dietz, Thomas, Elinor Ostrom, and Paul C. Stern
2003   The Struggle to Govern the Commons. Science 302:1907–1912.

Epstein, P. R.
1992   Pestilence and Poverty—Historical Transitions and the Great Pandemics.
        American Journal of Preventive Medicine 8:263–265.
1999   Climate and Health. Science 285:347–348.
2000   The WTO, Globalization and a New World Order. Global Change and
        Human Health 1:41–43.

Epstein, P. R., E. Chivian, and K. Frith
2003   Emerging Diseases Threaten Conservation. Environmental Health
        Perspectives 111(10):A506–A507.

Esrey, S. A., J. B. Potash, L. Roberts, and C. Shiff
1991   Effects of Improved Water Supply and Sanitation on Ascariasis,
        Diarrhoea, Dracunculiasis, Hookworm Infection, Schistosomiasis,
        and Trachoma. Bulletin of the World Health Organization 69:609–621.

Gladwell, Malcolm
2002   The Tipping Point: How Little Things Can Make a Big Difference.
        New York: Back Bay Books.

Hochschild, Adam
1998   King Leopold's Ghost. New York: Houghton Mifflin.

Hoerling, Martin P., James W. Hurrell, and Taiyi Xu
2001   Tropical Origins for Recent North Atlantic Climate Change. Science
        292:90–92.

Houghton, J. T., Y. Ding, D. J. Griggs, M. Noguer, P. J. van der Linden, X. Dai,
   K. Maskell, and C. A. Johnson, eds.
2001   Climate Change 2001: The Scientific Basis. Contributions of Working
        Group I to the Third Assessment Report of the Intergovernmental Panel on
        Climate Change. Cambridge: Cambridge University Press.

Hurrell, James W., Yochanan Kushnir, and Martin Visbeck
2001   The North Atlantic Oscillation. Science 291:603–605.

Kohn, Jorg
2004   The Political Economy of Sustainability: Towards the Integration of
        Economics, Social and Environmental Factors. London: Edward Elgar.

Lee, Jong-wook
2003   Global Health Improvement and WHO: Shaping the Future. Lancet
        362:2083–2088.

Levitus, Sydney, John I. Antonov, Timothy P. Boyer, and Cathy Stephens
2000   Warming of the World Ocean. Science 287:2225–2229.

Loff, Bebe
2002   World Trade Organization Wrestles with Access to Cheap Drugs Solution.
        Lancet 360:1670.

Lynch, J. W., G. A. Kaplan, and J. T. Salonen
1997   Why Do Poor People Behave Poorly? Variation in Adult Health Behaviours
        and Psychosocial Characteristics by Stages of the Socioeconomic
        Lifecourse. Social Science and Medicine 44(6):809–819.

MacMillan, Margaret
2003   Paris 1919. New York: Random House.

National Research Council
2002   Abrupt Climate Change: Inevitable Surprises. Washington, DC: National
        Academy Press.

Patz, Jonathon, Paul Epstein, Thomas Burke, and John Balbus
1996   Global Climate Change and Emerging Infectious Diseases. Journal of the
        American Medical Association 275:217–223.

Pollock, Allyson M., and David Price

2000   Rewriting the Regulations: How the World Trade Organisation Could
        Accelerate Privatisation in Health-Care Systems. Lancet 356:1995–2000.

Rosenzweig, Cynthia, Ana Iglesias, X. B. Yang, Paul R. Epstein, and Eric Chivian

2001   Climate Change and Extreme Weather Events: Implications for Food
        Production, Plant Diseases and Pests. Global Change and Human Health
        2:90–104.

Stephens, Carolyn, Simon Lewin, Giovanni Leonardi, Miguel San Sebastian Chasco,
        and Richard Shaw

2000   Health, Sustainability and Equity. Global Change and Human Health
        1(1):44–58.

Tong, Shilu, Rod Gerber, Rodney Wolff, and Kenneth Verrall

2002   Population Health, Environment and Economic Development. Global
        Change and Human Health 3(1):36–41.

United Nations

2002a   The Millennium Development Goals and the United Nations Role.
        Electronic document, www.un.org/millenniumgoals/MDGs-FACTSHEET1
        .pdf, accessed January 26, 2004.

2002b   Report of the World Summit on Sustainable Development: Johannesburg,
        South Africa, 26 August–4 September 2002. Electronic document,
        http://ods-dds-ny.un.org/doc/UNDOC/GEN/N02/636/93/PDF/
        N0263693.pdf?OpenElement, accessed January 26, 2004.

United Nations Development Programme

1997   Human Development Report 1997. New York: Oxford University Press.
1999   Human Development Report 1999. New York: Oxford University Press.

Victora, C. G., J. P. Vaughan, G. C. Barros, A. C . Silva, and E. Tomasi

2000   Explaining Trends in Inequities: Evidence from Brazilian Child Health
        Studies. Lancet 356(9235):1093–1098.

Watson, R. T., and A. J. McMichael

2001   Global Climate Change—the Latest Assessment: Does Global Warming
        Warrant a Health Warning? Global Change and Human Health 2:64–75.

Werner, D.

1998   The Struggle for Health: From Local to Global Level. Health Millions
        24:21–24.

World Health Organization
1997   Health and Environment in Sustainable Development. Geneva: World
          Health Organization.
2001   Macroeconomics and Health: Investing in Health for Economic
          Development: Report of the Commission on Macroeconomics and Health.
          Geneva: World Health Organization.

World Meteorological Organization
2003   Water and Climate. World Climate News, 23, June.

World Trade Organization
2004   The WTO in Brief. Electronic document, www.wto.org/english/thewto_e/
          whatis_e/inbrief_e/inbr00_e.htm, accessed December 20, 2003.

Index

About the Authors

Mary Anne Alabanza Akers is an associate professor at the University of Georgia's College of the Environment and Design. She has graduate degrees in urban and regional planning from the University of the Philippines and Michigan State University. Dr. Akers's research has focused on microenterprise development, community-based revitalization, and participatory design, and she has rendered numerous publications in the areas of community revitalization and socioeconomic and cultural aspects of design. She teaches courses in city planning, community theory and principles, and the sociocultural aspects of environmental design. She is now involved in research focusing on the relationship between environmental design and health.

Timothy Akers is the assistant dean for research and graduate studies in the WellStar College of Health and Human Services, Kennesaw State University, Kennesaw, Georgia. He holds a faculty position as professor of human services and serves as the first assistant dean for research and graduate studies in the university. Prior to this appointment, Dr. Akers was with the U.S. Centers for Disease Control and Prevention, where he served as the senior behavioral scientist and senior project officer in the Office of Minority Health; in the Office of the Director; and in the National Center for HIV, STD, and TB Prevention. Dr. Akers holds a joint bachelor of science degree in criminology and criminal justice from Metropolitan State College, Denver, Colorado; a joint master of science degree in urban studies and criminal justice from Michigan State

University; and joint doctorates in resource development (environmental science) and urban studies from Michigan State University.

George J. Armelagos is professor of anthropology at Emory University. Dr. Armelagos is particularly interested in the interaction of biological and cultural systems as applied to evolutionary problems. A major focus of his research has been on the evolution of food choice and on documenting the impact of agricultural development on health and disease. He has also written on the problems associated with the biological concept of "race." Professor Armelagos has authored a multitude of books, articles, and book chapters in the fields of medical and physical anthropology.

David G. Casagrande is an assistant professor of environmental anthropology at Western Illinois University. He received a master's degree in ecology and conservation policy from Yale University and a doctorate in ecological anthropology from the University of Georgia. His research is policy oriented and takes an ecological approach for explaining human behavior and cognition as interactions with natural environments. His current research interests include human environmental perceptions, information ecology of human ecosystems, urban ecology, and human adaptation. He has conducted research in Venezuela, Puerto Rico, Mexico, and postindustrial inner cities of the United States. Dr. Casagrande has published several research and policy articles in professional journals. He was coeditor in chief of the *Journal of Ecological Anthropology* from 2000 to 2003.

Nicole Consitt is a research assistant under the directorship of Dr. John Eyles at the McMaster Institute of Environment and Health. She is a recent graduate of the bachelor of science Health Studies program at the University of Waterloo.

Paul R. Epstein is associate director of the Center for Health and the Global Environment at Harvard Medical School and is a medical doctor trained in tropical public health. He has worked in medical, teaching, and research capacities in Africa, Asia, and Latin America and, in 1993, coordinated an eight-part series on Health and Climate Change for the British medical journal *Lancet*. He has worked with the Intergovernmental Panel on Climate Change (IPCC), the National Academy of Sciences, the National Oceanic and Atmospheric Administration (NOAA), and the National Aeronautics and Space Administration (NASA) to assess the health impacts of climate change and develop health applications of climate forecasting and remote sensing. Dr. Epstein is currently coordinating an international project *Climate Change Futures: Health, Ecological and Economic Dimensions,* in coordination with the Swiss Reinsurance Company and the United Nations Development Programme. This project involves scientists, UN agencies, NGOs, and corporate/financial sector leaders in the assessment of the new risks and opportunities presented by a changing climate.

John Eyles is a professor of geography at McMaster University, with cross-appointments in the Department of Clinical Epidemiology and the Department of Biostatistics, Sociology, and the Centre of Health Economics and Policy Analysis. He is author or coauthor of some two hundred books, peer-reviewed journal articles, and technical reports in the health and social sciences fields. He has carried out work for national and provincial organizations and governments in Canada. For example, he has worked with the International Joint Commission, the Ontario Ministry of Environment and Ministry of Health, the Prince Edward Island Department of Health and Social Services, Ontario Hydro, and the Canadian Chlorine Coordinating Committee. He serves on several expert panels and advisory committees and boards. His main research interests lie at the interfaces of environment, health, science, and policy. Dr. Eyles is also a fellow of the Royal Society of Canada and the director of the McMaster Institute of Environment and Health.

Greg Guest is a senior research associate at Family Health International, where he researches and publishes on numerous international public health topics. He received his doctorate in ecological anthropology from the University of Georgia and his master of arts in anthropology from the University of Calgary. He has carried out research in Mexico and Ecuador and currently manages several multisite projects in Africa. Dr. Guest has published in various public health and ecologically oriented journals, in addition to chapters in several edited volumes. His most recent work deals with HIV/AIDS prevention and behavioral aspects of clinical trials. Dr. Guest also has an ongoing interest in the integration of qualitative and quantitative methodology, as well as the translation of social science research to practice. He is currently editing a handbook for team-based qualitative research (with Kate MacQueen).

Kristin N. Harper is a doctoral graduate student in the Population Biology, Ecology, and Evolution Department at Emory University. She is especially interested in how social factors affect the evolution and transmission of infectious diseases and, in turn, how diseases affect human society. Kristin's work is supported by a predoctoral fellowship from the Howard Hughes Medical Institute.

Beverly Hill is working toward a doctorate in applied anthropology at the University of South Florida, where for the last two years she has served as the coordinator of wellness leadership. Her research interests include the political economy of health, ecological sustainability, and health policy analysis as a means of reducing health disparities and preserving human rights. She takes a special interest in the biocultural aspects of health and is preparing for research in the Pacific Islands (Micronesia) and Latin America. She currently is an evaluator for the state of Florida's Closing the Gap project, to aid community-based health programs in ending health disparities within the populations they serve. Additionally, her career trajectory includes working in citizen organizing, health advocacy, and political activism at the local and international levels.

Eric C. Jones earned his doctorate in environmental and ecological anthropology at the University of Georgia. His broad academic interests concern understanding the role of economic and environmental change on human demographics and human social relations. His focus on human mobility as a response to pressures on broad and fine scales has led him to research pioneer colonization, rural-rural migration, rural-urban migration, and international labor migration. Dr. Jones is a postdoctoral researcher at the University of North Carolina, Greensboro, studying disaster response, particularly changes in social relations. Before working at the university, Dr. Jones was a faculty member for a comparative international study-abroad program at Boston University, where he taught undergraduate students about ecology and globalization.

Suzanne E. Joseph is an assistant professor of anthropology at the University of Massachusetts, Dartmouth. She received her doctorate in anthropology from the University of Georgia. She has also held an Andrew W. Mellon Postdoctoral Fellowship in Interdisciplinary Approaches to Microdemography at Brandeis University, where she was hosted by the Departments of Anthropology, Sociology, Human Biology, and Environmental Studies. A major focus of her research is to document and understand the demographic and health implications of inequality in time and space. She has conducted research in rural parts of Lebanon and Syria. Dr. Joseph is presently working on a book based on her dissertation research.

Thomas L. Leatherman received his doctorate in anthropology from the University of Massachusetts and his master of arts in anthropology from the University of Arkansas. He is professor and chair of anthropology at the University of South Carolina. Specializing in biocultural theory and practice and the political ecology of health, Dr. Leatherman has authored numerous book chapters and has published in journals such as *American Journal of Physical Anthropology, Human Organization,* and *Social Science and Medicine.* He is also coeditor of *Building a New Biocultural Synthesis: Political–Economic Perspectives on Human Biology* (1998). Dr. Leatherman currently serves on the editorial board for *Medical Anthropology Quarterly.*

George E. Luber is an epidemiologist and medical anthropologist with the National Center for Environmental Health at the Centers for Disease Control and Prevention. He received his doctorate in ecological anthropology from the University of Georgia and a master's degree in applied anthropology from Northern Arizona University. His doctoral work in Chiapas and Oaxaca, Mexico, focused on the biocultural epidemiology of "second hair" illness in two Mesoamerican societies: the Tzeltal and Mixe. A summary of findings from this research is forthcoming in *Nutritional Anthropology.* In addition to working with the epidemiology of folk illnesses and infant nutrition, Dr. Luber has conducted research on HIV vaccine trials, pesticide exposure, heat waves, and ciguatera marine toxins.

Ann McElroy received her doctorate in anthropology from the University of North Carolina, Chapel Hill. She teaches medical and applied anthropology at the State University of New York at Buffalo, where she is codirector of the applied medical anthropology concentration for master's students. She has done research on the impact of culture change and modernization on Inuit children's identity in the Canadian Arctic for 37 years. In recent years she has been interviewing elderly Inuit about their memories of the early 20th century during the transition between living on the land and living in settlements. Other research interests include migrant worker health, anthropology and disability, and environment and health. Dr. McElroy is coauthor of the popular textbook *Medical Anthropology in Ecological Perspective* (fourth edition, 2004).

Linda M. Whiteford is a professor of anthropology and chair of the Anthropology Department at the University of South Florida. She received her doctorate in anthropology from the University of Wisconsin and her master's in public health from the University of Texas. She has conducted research in Cuba, the Dominican Republic, Nicaragua, Costa Rica, Ecuador, and Bolivia. For the last ten years, Dr. Whiteford has been working on infectious and communicable diseases such as cholera, dengue fever, and diarrhea and on health system policy and analysis. Currently she is combining her interest in vector- and water-borne diseases with health consequences of forced evacuation or other population movements. Dr. Whiteford's most recent book (with Lenore Manderson) is *Global Health Policy, Local Realities: The Fallacy of the Level Playing Field.*